Dear Papa/William,

Happy 101st Birthday!

Hope this book brings you enjoyment.

Love,
Bronwen and Tim

Dear Papa (William)

Happy 101st

Birthday!

Hope this book

brings you

enjoyment.

Love,

Bronwen and Tim

TRAINS

The World's Most Scenic Routes

Publications International, Ltd.

Written by: Ian Feigle

Images from: Shutterstock.com and Wikimedia Commons

Louis Weber, CEO
Publications International, Ltd.
8140 Lehigh Avenue
Morton Grove, IL 60053

ISBN: 978-1-64030-652-3

Manufactured in China.

8 7 6 5 4 3 2 1

CONTENTS

INTRODUCTION

Ride the rails with *Trains* as it traverses the world to showcase the groundbreaking routes that have come to connect our modern age. Traveling through history along passenger and freight services that have evolved from horse-drawn rail carriages to steam, diesel, and electric powered locomotives, *Trains* covers some of the most intriguing and scenic routes from railway history.

There is perhaps no better example of the industriousness and fortitude of humanity as that of the railway. Hundreds of thousands of miles of track have been laid through dense, old-growth forests, tenuous river valleys and gorges, and the world's most drastic and difficult terrain with the brute force and drive of the human will. Tunnels have been bored through alpine peaks and around glaciers. Trestle bridges and viaducts have been erected over bogs, permafrost, and canyons. Trains and their railways have proven to be an indispensable technology that is not only a remaining vestige of industrialization, but a beacon for us to follow in order to meet our growing and changing transportation needs.

Passenger service on historic and heritage tourist railways has grown in the past few decades. As railway construction boomed through out the world in the mid-nineteenth and twentieth centuries, many of the shorter, older, and privately owned railways were either nationalized and closed or forced out of business due to shrinking freight demand. But as the ownership and operation of many railways switched hands between railway companies and governments through the years, the historic and scenic routes that once thrived have not been forgotten.

North American routes like the White Pass & Yukon Route on the United States and Canadian border, the Grand Canyon Railroad in Arizona, and the nation-spanning California Zephyr have all faced phases of heavy use or near abandonment, but are thriving tourist attractions today. European mountain railways that ascend into the niches of the Alps, like the Bernina Express of Switzerland, the Semmering Railway of Austria, and the Bavarian Zugspitze Railway in Germany, have all had thriving passenger services since the early twentieth century. *Trains* covers these routes and many more to provide a glimpse into the vast and intriguing history of the Industrial Revolution's preferred method of transport and the various ways it has been revitalized since.

The historic Main Line of the Trans-Siberian Railway was built around the southwestern shore of Lake Baikal, the largest freshwater lake by volume in the world, along with almost two hundred bridges and thirty-three tunnels. The railway replaced the ferry service that traditionally ran cargo from one side of the lake to the next.

THE TRANS-SIBERIAN RAILWAY

The Trans-Siberian Railway is a transcontinental railway that spans Russia, from the capital of Moscow to the far-eastern administrative center of Vladivostok. The 6,888 miles of track that connect hundreds of large and small cities across Russia, Ukraine, Siberia, North Korea, Mongolia, and China make it the longest railway in the world. The Trans-Siberian Line, the main historic line of the railway, was built between 1891 and 1916 under the impetus of Tsar Alexander II, while auxiliary lines were built later and are still expanding into new territory today. The railway is so long that it crosses over eight different time zones, but it takes only six days to travel from end to end.

Before the railway existed, transportation to and from Siberia was heavily restricted because of the region's brutal winters. Other than the Great Siberian Route, a historic highway that connected European Russia with Siberia and China, there were very few suitable options to transport goods to and from the region before the railway was finished.

Today the railway serves as a main line for passengers and cargo to travel across the vast land of Russia. Nearly thirty percent of all of Russia's exports travel on the line, and cargo coming from Beijing can make it to Hamburg, Germany, in fifteen days—a much shorter shipment period than ocean liners can offer. The importance of the railway to the Russian economy cannot be overstated, and the Russian government continues to shorten the time it takes to make it across the country, while also maximizing the amount of cargo each train can undertake.

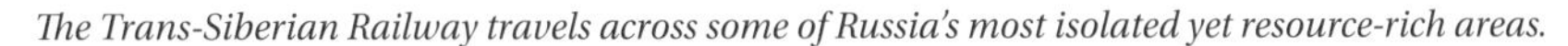

The Trans-Siberian Railway travels across some of Russia's most isolated yet resource-rich areas.

A train running along the Olkha River.

THE KURANDA SCENIC RAILWAY

It takes an hour and fifty-five minutes to reach the Kuranda Train Station from the Cairns Train Station.

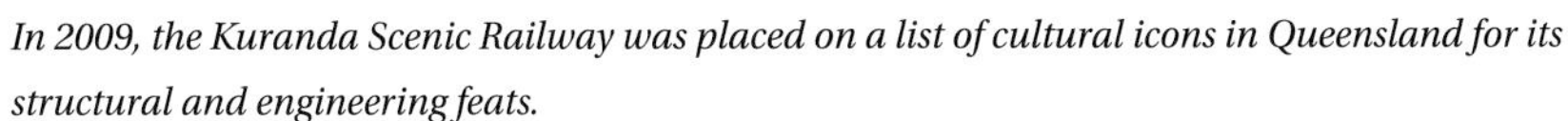

In 2009, the Kuranda Scenic Railway was placed on a list of cultural icons in Queensland for its structural and engineering feats.

Barron Falls diving from the mountains toward sea level.

The Kuranda Scenic Railway is a tourist railway that snakes its way through the Barron Gorge National Park in Queensland, Australia, along the Barron River. Ascending up the Macalister Mountain Range, the twenty-three mile long railway was originally built to ship goods and mining supplies between Cairns and the resource-rich area of Kuranda. Construction finished in 1891, and the first tourist train ran in 1936. Since then tourists have continued to flock to this railway to experience the Wet Tropics Rainforest, rushing waterfalls, and the area's unique flora and fauna. Along the route, passengers will see numerous waterfalls including Barron Falls and Stoney Creek Falls, cross over fifty-five bridges, and pass through fifteen tunnels.

The railway's five-year construction period was arduous and dangerous, but much needed. Tin miners in the late nineteenth century faced famine conditions and found it extremely difficult to make passage from the coast into the mountains to resupply their camps. Years of surveying and logistical disputes between municipalities finally led to the decision of the Barron Gorge's route from Cairns to Kuranda. Many lives were lost as crews cleared dense forests, working their way to the top of the Macalister Mountain Range. The gorge's slopes were angled at nearly forty-five degrees and the earth was covered in several feet of rock, dead vegetation, and mud—all adding to the difficulties in completing the railway. A majority of the clearing was done by hand, with buckets and shovels used as the main tools of excavation. In the end, 812,000 tons of land was removed from the area.

Despite all the work, the Kuranda Scenic Railway has not made the Wet Tropics Rainforest any more manageable. The tracks were damaged in 1995 after a severe rockslide, and the tracks were closed for three months in 2010 after a mudslide derailed a train and injured a handful of the 250 passengers. The train runs twice a day with a variety of class tiers to enhance your trip.

Right: *The Wet Tropics Rainforest that surrounds the Barron Gorge has some of the most unique and endemic flora in the world, including nearly ninety species of orchids.*

The track of the White Pass & Yukon Route is a narrow-gauge track at three feet in width. The narrow gauge allowed for construction costs to be much lower than standard gauge because the smaller roadbed required less blasting.

THE WHITE PASS & YUKON ROUTE

The White Pass & Yukon Route is a vestige of the determination that fueled prospectors during the Klondike Gold Rush of 1897. Built in notoriously rough and mountainous terrain along the U.S./Canada border, the White Pass & Yukon Route runs from Skagway, Alaska, through British Columbia, Canada, and then terminates at Carcross, in the Yukon Territory of Canada. Although the original train route made it as far north as Whitehorse, Yukon Territory, the tracks that lead to Whitehorse are no longer certified for use by the Canadian Transportation Agency.

Before the railway was built, Skagway was only accessible by sea, and the gold fields in the Yukon Territory were only accessible from Skagway by traversing over the White Pass or the Chinook Trail, the only two passes through the Boundary Range of the Coastal Mountains, by foot. In those days, prospectors coming from America were often told to turn around by the Canadian authorities at the border if they did not have adequate supplies for the winter—often considered a ton's worth of goods, requiring multiple trips up and down the pass. The Klondike Gold Rush began in 1897, a year after George Carmack and Skookim Jim's discovery of gold in Bonanza Creek, and the construction for the White Pass & Yukon Route started in 1898 with the help of various companies and British investors.

By time the railway had opened, the gold rush was bust, and many of the near hundred-thousand people who came running into the treacherous territory began to retreat home. The railway was used to haul copper, silver, lead, passengers, and other freight through the largely inaccessible areas of Alaska and Yukon Territory until 1982. As Alaskan tourism began to increase in the late twentieth century, the prospect of using the railway as a tourist attraction became more viable. Cruise ships often visited the port of Skagway, and the tracks of the White Pass & Yukon Route lie right at the edge of the port, creating a convenient way for passengers to gain access to the Alaskan wilderness. Today the railway operates as a tourist attraction, leading visitors into the vast wilderness of Alaska.

Tours on the White Pass & Yukon Route are available between May and October.

The Boundary Range contains many ice fields and acts as a physical marker for the border between southeast Alaska and British Columbia, Canada.

THE FERROCARRIL SANTA ANA

There are a variety of services operating along the Ferrocarril Santa Ana, including luxury trips on the Belmond Hiram Bingham train, which is named after the American discoverer of Machu Picchu, and more frugal trips for those who are more adventurous.

This Peruvian route takes you to the ruins of the fifteenth century Incan citadel of Machu Picchu along the PeruRail system. PeruRail is divided into two sections: the Ferrocarril Santa Ana, which terminates at the Machu Picchu station of Aguas Calientes, and the Ferrocarril del Sur, which travels from the southern coast of Peru to the historic Incan capital city of Cusco. Founded in 1999 by a Peruvian entrepreneur, the PeruRail system is the third highest railway in the world following the Qinghai-Tibet Railway in China and the FCCA (Ferrocarril Central Andino), which is also in Peru.

Right: *The Ferrocarril Santa Ana runs along the Upper Urubamba River in the Sacred Valley of the Incas. This section of the river has many centuries of history, very well developed irrigation systems, and a high density of population around it.*

Bottom: *The station at Aguas Calientes at the foot of Machu Picchu.*

Ferrocarril Santa Ana's original route left from Cusco up a steep incline into the mountains, known as "The Zig-Zag", but today, trains leave out of the Poroy Train Station, which is just outside of the city limits of Cusco. (No trains currently go through the city of Cusco.) From Poroy, the train enters the Sacred Valley of the Incas and travels northwest along the Urubamba River through the Area de Conservacion Regional Choquequirao, a region of historic importance that contains many sites of Incan heritage, to the city of Aguas Calientes. The tracks from Aguas Calientes once continued out of the city farther into the jungle, but the tracks were recently damaged from flooding. From Aguas Calientes, Machu Picchu is just over three miles away, which can be reached by foot in about an hour and a half.

THE TRANZALPINE

The TranzAlpine is the most famous tourist train in the country of New Zealand, and perhaps the greatest train ride in the world as it spans northwest across New Zealand's southern island from Christchurch to Greymouth. The route carried just over 200,000 passengers a year in 2007, but those numbers have dropped a bit since the 2011 Christchurch earthquake that struck the area. Operated by Great Journeys of New Zealand, the TranzAlpine runs seven days a week and takes about five hours to run from end to end.

Various operators have owned the TranzAlpine over the years. It opened in 1987 as a passenger service on the Midland Line, which was used to transport freight between the Rolleston and Greymouth stops since 1923. Before 1923, the Midland connected the economic centers of Christchurch, Timaru, and Dunedin in the arable plains of the east side of the island, but the track was then expanded to provide transportation to the resource rich western part of the island. Connecting this original track to the western portion of the island was a tremendous and laborious task. New Zealand's

Southern Alps were directly in the proposed track's path, and hindered the completion of the track for many years. To get over the Southern Alps, men with picks and shovels in hand cleared the roadbed. All in all, they built sixteen tunnels and four major viaducts to complete the task.

This treacherous and mountainous terrain is what gives the TranzAlpine its majestic views. Departing from Christchurch, the TranzAlpine crosses the fertile Canterbury Plains, the largest flat tract of land in New Zealand, and passes the highly braided Waimakariri River. From there it continues west into the Southern Alps and over the Waimakariri River Gorge. Passing through the five-mile long Ōtira Tunnel, the train overcomes Arthur's Pass and begins its descent into New Zealand's wild and sparsely populated West Coast.

Right: *The station at Arthur's Pass (2,425 feet) is just east of the Otira Tunnel, which is 5.3 miles long and ranks as the seventh longest tunnel in the world.*

Above: *A view of the Southern Alps from the TranzAlpine train. The Southern Alps were named by Captain James Cook in 1770.*

Below: *The alpine and grassland habitats of the Southern Alps' higher elevations hold a quarter of New Zealand's plant species.*

THE TALYLLYN RAILWAY

Locomotive No. 4, the Edward Thomas, was bought in 1951 from the neighboring Corris Railway in order to resume operations while repairs were undertaken on the original two locomotives, the Talyllyn and the Dolgoch.

As the first narrow gauge railway in Britain to be authorized by an Act of Parliament and the first heritage railway in the world to be preserved by volunteers, the Talyllyn Railway holds a special place in Welsh history. This relatively short and narrow railway transports tourists for about seven miles between the mid-Wales coastal town of Tywyn to the Nant Gwernol Station outside the village of Abergynolwyn. The original terminus of the rail was in Abergynolwyn until 1976 when an old mineral line was used to extend the service to Nant Gwernol. This single-track railway has become so popular with tourists that the "one engine in steam" policy had to be abandoned in order to keep up with demand, which required passing loops and more stringent single-line control in order to prevent collisions.

Originally conceived as a way to freight slate from a nearby quarry at Bryn Eglwys to the standard gauge Aberystwyth & Welsh Coast Railway that ran to Tywyn, the line opened in 1866 to provide both freight and passenger services. The railway ran successfully for many decades with just two locomotives transporting passengers and slate, but by the turn of the twentieth century, it became clear that the fate of the railway heavily depended on the operations of the quarry—the main employer of the Abergynolwyn area. By the time the quarry's lease expired in 1910, no one was willing to renew the lease for the overburdened land. But in an act of desperation, local politician and businessman Henry Haydn Jones bought the quarry and railway to keep the economic situation in the town stable.

The Talyllyn Railway and quarry barely hung from a thread for the next few decades. Being one of the few railways not included in the nationalization of the British railway system in 1947, the Talyllyn Railway was left in an even more tenuous economic position. The railway was saved from abandonment by a committee of volunteers who petitioned to acquire the rights of the railway from the estate of Jones. In 1951, the committee began operating the railroad as a tourist attraction. Although preservation was difficult due to all of the restoration the rolling stock and line needed, the Talyllyn Railway has successfully preserved the rail history of the Abergynolwyn area.

Top left: *The original gauge of the Talyllyn Railway was two feet and three inches wide, which is an unusual gauge of track, requiring locomotives to be modified in order to operate on the track.*

Bottom left: *The Talyllyn Railway steaming through the Welsh countryside.*

THE ROCKY MOUNTAINEER

In 1885, the last spike was driven into the tracks that connected Canada's east and west coasts, completing 3,000 miles of the transcontinental railway that would unite the country divided by the Northern Rocky Mountains. Operated by Canada's Via Rail system until government subsidies were cut drastically in 1990, the rights to the tracks and rolling stock that wind through the rainforests of British Columbia and around the towering peaks of the Rocky Mountains were sold to the Rocky Mountaineer Railway Company. And now, traveling in the Rocky Mountains is more luxurious than ever.

The Rocky Mountaineer Railway Company operates on four different routes through British Columbia, Alberta, and Washington State. The "First Passage to the West" route is the only passenger service that runs on a section of the tracks that originally connected the country, the Canadian Pacific Railway, more than 130 years ago, traveling between Banff National Park and Vancouver along the Kicking Horse River. The "Journey Through the Clouds" route leaves from Vancouver, through Kamloops, and then up to Jasper National Park, passing the Albreda Glacier, Pyramid Falls, Fraser Canyon, and Mount Robson, the highest peak in Canada.

The "Rainforest to Gold Rush" route is a three-day trip through the varied ecosystems of the region, including the coastal rainforests of Vancouver, the Rocky Mountain Trench, and the pastures of Cariboo Plateau. The "Coastal Passage" has connected Seattle to Vancouver to allow for a continuous train ride up from the states into Canada's richest natural wonders.

With The Rocky Mountaineer, you can travel anywhere between one to fourteen nights through the Canadian wilderness, depending on the route you would like to take. There are three levels of service in which you can book your trip with too, from RedLeaf—the most basic experience—to GoldLeaf—the most elaborate and sophisticated passenger experience.

Top: *The GoldLeaf service provides a travel experience like none other in a custom-designed, glass domed two-level coach.*

Above: *The Rocky Mountaineer provides service between April and mid-October, departing from the east- and west-bound stations multiple times a week.*

Left: *The Rocky Mountaineer is the busiest privately owned train service in North America.*

THE GRAND CANYON RAILWAY

Although train service was once discontinued because of the popularity of cars and falling ridership, today the Grand Canyon Railway keeps nearly 50,000 cars away from this natural wonder every year.

Passenger service to the Grand Canyon hasn't always been consistent since the Grand Canyon Railway opened in 1901. Built by the Santa Fe Railway as a branch between Williams, Arizona, and the Grand Canyon Village, the Grand Canyon Railway offered tourists a reliable way to visit this natural wonder hidden in the high desert of northern Arizona for less than four dollars. But tourists weren't the main driver for building the railway at first. Investors wanted a way to transport silver, lead, copper, and gold ore to the Santa Fe Railway—which connected the country by rail from Chicago to Los Angeles—from the Anita Mines just forty miles north of Williams. But as mining at the site proved unsustainable, investors were still able to make a return on their money through tourism.

With the growing popularity and ubiquity of the automobile in the mid-twentieth century, train services throughout the nation were struck with shrinking passenger numbers, and the Grand Canyon Railway was no exception. The train's original stint of passenger service lasted until 1968, while the Santa Fe Railway continued to use the route as a freight service until 1974. But the route was completely abandoned by 1977, with a few unsuccessful attempts to resurrect the service to no avail. But by 1988, the continued operation of the railway was beginning to look hopeful when the Biegert family acquired the railway. By 1989, the Biegerts had the railway carrying passengers again to the South Rim of the Grand Canyon and continued to do so until 2006 when Xanterra Parks & Resorts bought the railway. Under Xanterra ownership, the traditional steam engines were no longer used in light of the growing energy crisis of the twenty-first century. Today, diesel engines make a majority of the runs, but engines that run on vegetable oil can be seen on the tracks from time to time.

On the sixty-four mile trip from Williams to the Grand Canyon, you will float through the high desert of Arizona, thousands of feet above sea level. You will traverse out of the Coconino National Forest, filled with evergreen Ponderosa Pines, up to the Kaibab National Forest that lines the South Rim with juniper and sagebrush. To the east during your trip, you will see the San Francisco Peaks, the highest point in the state of Arizona, which lie just outside of Flagstaff, Arizona. Today, the entire area is completely serene compared to the seismic, volcanic, and geographic forces that have so violently shaped this majestic portion of the American Southwest.

Nearly 4,000,000 people visit the Grand Canyon every year, so be sure to plan ahead when trying to visit the nation's busiest national park.

Young elk near the Grand Canyon. The Kaibab National Forest is home to various types of mammals including elk, mule deer, white-tailed deer, pronghorn, coyote, turkey, cougars, bobcats, and black bears.

THE WEST HIGHLAND RAILWAY

Left: **Left:** *The Glenfinnan Viaduct is a mass concrete structure that is an icon of the West Highland area. The viaduct has been featured in many films including four Harry Potter films, which has led some fans to climb the viaduct and nearly get hit by passing trains.*

Travelling through the mountainous region of Scotland's Highlands, the West Highland Railway travels from Scotland's most densely populated city of Glasgow northwest through one of Europe's most sparsely populated areas near the port cities of Oban and Mallaig on the west coast. Constructed in phases that slowly connected the difficult and loch-ridden area of the Highlands with the economic center of Glasgow, the last branch of the railway, Craigendoran to Mallaig, was finished in 1901 after twelve years of construction.

Breaking through the sod in 1889 for this last branch, the construction crew faced difficulties in making substantial progress on the line because of the area's sheer remoteness. Basic roadways were not available to ship construction equipment into the area and the soil out of the area. But despite the slow progress during its construction, the route has become one of Europe's most cherished railways for the beautiful scenery the line passes through.

Although trains are generally used for their expediency over other forms of intercity travel in many parts of the world, the West Highland Railway can take a much longer time from one destination to the next than it would take by car. The reason for this is the single-track design the railway was constructed with, allowing for only one train at a time to pass as other trains wait at stations with crossing loops. On the journey, you will cross over numerous lochs, including Loch Lomond, Long, and Treig, stop at Europe's remotest station in Corrour, which is not accessible by any road, and also pass over the Glenfinnan Viaduct, a mass concrete structure that overlooks the Glenfinnan Monument.

A sleeper train crossing the Rannoch Viaduct over the boggy peat expanses of Rannoch Moor.

A steam engine arriving at Corrour Station, a train station not accessible by public roads.

THE DURANGO & SILVERTON NARROW GAUGE RAILROAD

The narrow gauge of track that the D&SNG is named after is only three feet wide, one foot and eight and a half inches narrower than today's standard railroad gauge.

A relic from the mining days of the Wild West, the Durango & Silverton Narrow Gauge Railroad (D&SNG) has been in operation for over 130 years, although it carries more tourists today than the traditional silver and gold ore it was built for. The portion of track that is today known as the D&SNG was originally a part of the San Juan Extension of the Denver & Rio Grande Railway (D&RG). From Denver, the San Juan Extension traveled south to Santa Fe, New Mexico, and then west toward Durango. The Silverton Branch of the San Juan Extension then went north from Durango toward the abundant mining areas in the San Juan Mountains around Silverton, which is the same track the D&SNG operates as today.

The tracks were laid into Durango on August 5, 1881, and then expanded into Silverton by July 10, 1882. It was not long before the Silverton Branch was hit hard by economic factors. The Panic of 1893, the oversupply of silver, the decreased mining ventures in the area, and declining revenue from passengers all brought the Silverton Branch's service nearly to a halt. And with all economic factors aside, environmental factors, like mudslides and flooding, were also huge hurdles that nearly stopped the railway in its tracks.

Since the 1980s, the D&SNG has been bought and sold by various private owners, and it continues to be the one of only a few rail services that still operate steam locomotives in the nation. It is also only one of two remaining narrow gauge tracks still operating in what was once Colorado's very extensive narrow-gauge network. It is a federally designated National Historic Landmark and a Historic Civil Engineering Landmark designated by the American Society of Civil Engineers.

Bottom left: *The D&SNG is forty-five miles long and takes three hours to go from one end to the next.*

Bottom right: *The D&SNG crosses the Animas River five times in the forty-five mile stretch between Silverton and Durango.*

THE GHAN

It takes fifty-four hours to travel the 1,851 miles between Darwin and Adelaide.

Abave: *Before diesel engines were available, the use of steam locomotives through the arid outback of Australia was a problem because many routes lacked an adequate supply of water.*

Below: *It took 126 years of planning and $1.3 billion to build the track between Alice Springs and Darwin.*

Traversing Australia's central territories, the Northern Territory and South Australia, the Ghan is a near 2,000 mile trip through Australia's harsh interior. Originally finished in 1929 as a narrow-gauge connection between Adelaide on the south coast to Alice Springs in the center of the continent, the Ghan has been finished in recent decades to cross the entire expanse of the continent. The construction began in 1878 from Port Augusta, slowly making it further north into drought ridden areas, with stops eventually making it to Hawker in 1880, Beltana in 1881, Marree in 1884, and Oodnadatta in 1891. At that time, a trip by camel was the only way to make it to Alice Springs from Oodnadatta. The railway made it to Alice Springs nearly thirty years later.

The railway followed the old telegraph route that ran through the center of the continent, which is believed to have been built on the route taken by John McDouall Stuart's 1862 north-south crossing of the Australian continent. Delays were a common occurrence for the railway because the tracks were often washed out from flash flooding in the area. In order to keep service somewhat regular, a flatcar that carried construction tools and sleepers was attached to the end of the train to allow for passengers to help with repairs when needed.

Today, a new standard-gauge track has replaced the old narrow-gauge track that originally ran from Adelaide to Alice Springs until 1980. Construction from Alice Springs to Darwin began in 2001 and was finished in 2004. It is considered the second-largest civil engineering project completed in the Australian continent. This connection to Darwin now allows for greater access to transportation for aboriginal peoples in the area, as well as creating the possibility of Darwin becoming a new trade center with a link to Asia. The Northern Territory is Australia's third largest municipal state, its least populated, and its least inhabitable due to its arid conditions, but the Ghan now offers a level of connection with the rest of the continent that the Northern Territory has never had before.

THE DENALI STAR

Trekking through more than 300 miles of Alaskan wilderness, the Denali Star provides seasonal rail service between Anchorage and Fairbanks along the Alaska Railway. From May to September, the Denali Star is an easy way for tourists to get to Denali National Park with dome cars for viewing the tremendous, snow-covered peaks of the Alaska Range. Daily service runs both directions toward Fairbanks and Anchorage, while the Aurora Winter Train operates for the rest of the year a few days out of the week on a flag-stop service.

***Right:** The dome cars allow passengers to see much of the expansive wilderness the train passes through.*

The Denali Star runs on the Alaska Railroad system, which was built by the United States government between 1909 and 1914. Aside from passengers who want to get a glimpse of the Alaskan wilderness, large amounts of freight makes its way into the interior by way of this rail. Although, the Alaska Railroad is not connected to the contiguous forty-eight states, freight is received from three rail barges that operate between Harbor Island, Washington, and Anchorage, Alaska. The railway also operates as a lifeline to residents who live in the Hurricane Area, which is not accessible by road. Residents of this area receive all of their food and goods by rail, and can enter or depart from the train by using the flag-stop service, which allows them to have the train stop anywhere they need without the presence of a depot or station. The Denali Star is as wild as a train service can get, taking passengers and residents of the Alaskan wilderness into truly remote areas that have no other access to the outside world.

***Top left:** Diesel engines have been operating on the railway since 1944, while steam engines continued to operate until 1966.*

***Bottom left:** The Aurora Winter Train provides semi-weekly service during the winter months.*

Locomotive #73 has been modified to burn a combination of natural gas and diesel as a cleaner fuel source. Locomotive #70 also underwent this conversion but has since been reverted back to strictly diesel.

Operating as a tourist expedition through northern California's wine-rich Napa Valley since 1989, the Napa Valley Wine Train has transported nearly two million passengers between the four stations along the line. Following much of California State Route 29, the railway runs between Napa and St. Helena, California, for some eighteen miles of the track's original forty-two mile length. Built in 1864 by California pioneer Samuel Brannan, the train originally ran from Vallejo, California, to help bring tourists into the resort town of Calistoga, California. Going bankrupt only years after it opened, Brannan's railway was bought by various other railway companies throughout the next one hundred years for use as freight transportation in the area.

The new wine-centric focus of the railway came about in 1989 after the Southern Pacific Railroad notified local officials that it planned to abandon operating on the line. At that time, the Napa Valley Railroad was formed by a group of local entrepreneurs in order to provide a train service for wine enthusiasts. Although the new service was strongly opposed by locals, the Napa Valley Railway was given right of way on the tracks and began operations September 16, 1989.

The tour is a three-hour journey through some of America's most luscious vineyards and expensive farmland, paired with a meal cooked on-site and local wines available in the lounge cars. There are optional stops for passengers to get off and take tours at any of the region's numerous wineries. Or you can simply enjoy the ride as you pass the time with a fine glass of wine.

Abave: *The Napa Valley Wine Train operates twice a day, one leaving at 11:30 am and the other at 6:30 pm.*

Below: *By the end of the nineteenth century, Napa Valley had as many as 140 wineries in the area.*

THE B&O RAILROAD

Shown here in the twentieth century, Ellicott City was the B&O's first terminus. The stone building dates from 1831, making it America's first railroad station. Today, it has been magnificently restored and is open as a museum.

The Baltimore & Ohio Railroad gets credit for commencing the first truly modern railroad, even though John Stevens had predated the B&O with what was for its time an audacious plan to build a railroad across Pennsylvania. The B&O originated in the minds of leading Baltimore merchants in 1826, and it was chartered in early 1827. Three years later, the B&O opened America's first common-carriage railroad (thirteen miles to Ellicott's Mills) as the first leg of a proposed 380-mile, double-tracked line over the Allegheny Mountains intended to tap into the burgeoning western traffic moving on the Ohio-Mississippi River system. This was different in scale, intent, and risk from the short railroads that were its contemporaries. So different, in fact, that the B&O was a tremendous gamble.

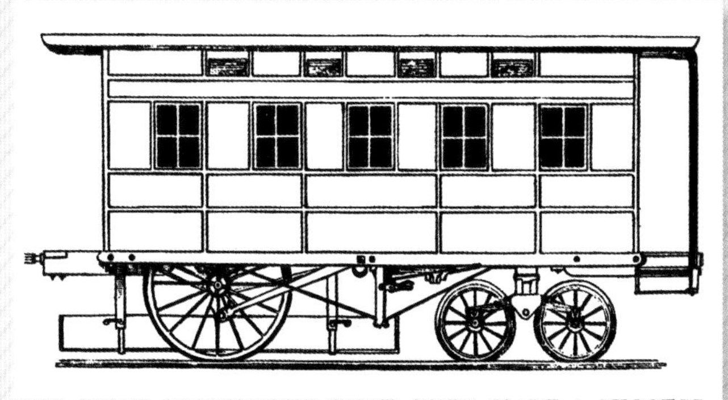

Above: *In 1851, Dr. Charles Grafton Page tested an experimental electric locomotive on the B&O's tracks out of Washington, D.C.*

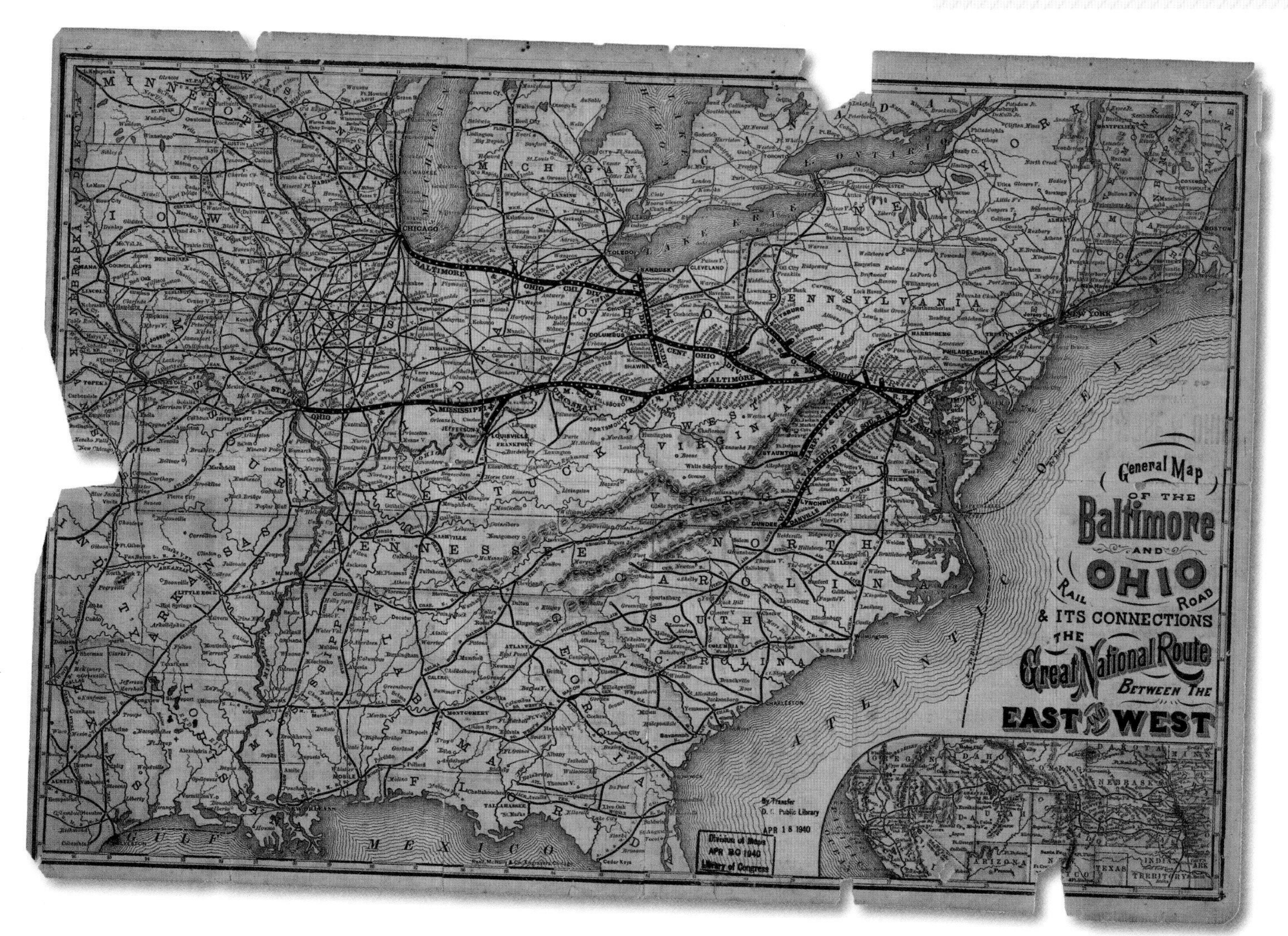

Left: *A map of the service routes provided by the B&O Railroad in 1876.*

The Alishan Forest Railway is fifty-three miles long, but that length depends on whether certain portions of the track are closed are not.

THE ALISHAN FOREST RAILWAY

Built by the Japanese Colonial Government in 1912 during their control of the island of Taiwan, the Alishan Forest Railway was constructed to promote logging in the area of Chiayi County, Taiwan. After the Treaty of Shimonoseki, Japan took control of Taiwan and soon discovered the rich resources of lumber in the country's forest. Construction on the narrow-gauge railway began in 1906 and took six years to complete. According to the website Rtaiwanr.com, the railway is considered to be one of the five wonders of the resort town of Alishan, along with the forest itself, the sea of clouds, the sunrise, and the sunset.

The train's original departure point into the Alishan National Forest and surrounding mountains was from Chiayi City toward the Zhaoping Station, but major sections of the railway tend to shutdown service because of its vulnerability to damage. Typhoons in the twenty-first century have closed major portions of the track for many years, while there is also a long history of landslides. As the train works from 100 feet above sea level to nearly 6,000 feet above sea level, the steep gradients begin once you leave Chiayi City.

With Z-shaped switchbacks, fifty tunnels, and seventy-seven wooden bridges, the Alishan Forest Railway is a feat of engineering that attracts crowds of tourists everyday. The numbers of commuting passengers on the railway has declined, making the railway primarily a tourist attraction for people to enjoy the fantastic views that it provides. But the views come with a price. Aside from the damage sustained by typhoons in the past few years, several incidents have occurred that have led to passenger fatalities. A tunnel collapsed in 1981 leading to nine deaths and thirteen injuries. In 2003, the train derailed near the Alishan Railway Station which resulted in 156 people being injured and seventeen people being killed. And again in 2011, the train derailed causing five deaths and 113 injuries.

The Alishan Forest Railway was operated by a private owner until 2008, before the train's service was handed over to the Forestry Bureau of Alishan. Diesel locomotives mainly operate on the tracks, but there are special expeditions that use steam locomotives on occasion for public events.

Top right: *Nearly 5,500 people ride the railway everyday.*

Center right: *Taiwan has tried to have the Alishan Forest Railway listed as a World Heritage Site, but because of Taiwan's exclusion from the United Nations, it is unlikely to be recognized as such.*

Bottom right: *The Mianyue Line on the Alishan Forest Railway has been closed indefinitely since 1999 after earthquakes struck the area.*

THE BLUE TRAIN

The Great Karoo region is defined by its arid climate, cloudless skies, and extreme cold and heat. The area is home to a humungous fossil record that gives us a picture of the diverse ecosystem that occupied this area hundreds of millions of years ago.

Left: *The Blue Train travelling along the coast of the Indian Ocean.*

Below: *The community of St. James Beach nestled between the tracks of the Blue Train and the waters of the Indian Ocean.*

Bottom: *The Kaaimans River Bridge outside of Victoria Bay on the Blue Train's Garden Route.*

Known as one of the most luxurious train rides in the world, the Blue Train of South Africa offers an exclusive experience with butler service, lounge and observation cars, soundproofed picture windows, carpeted compartments, and sleepers that feature bathrooms and even bathtubs. Traveling 990 miles between the administrative center of Pretoria to the coastal city of Cape Town, the main route of the Blue Train was originally built in 1923 to transport passengers to the port of Cape Town to book passage back to England. The Garden Route is the only other route that the train service provides, running along the coast of the Indian Ocean from Cape Town to Port Elizabeth. The Blue Train used to operate along two other routes until 2004; one from Pretoria to Kruger National Park in northeast South Africa, which closed due to low passenger rates, and another that ran from Pretoria to Zimbabwe's Victoria Falls, which closed because of often exorbitant ticket prices imposed on the rail network within the financially strapped country of Zimbabwe.

Service on the Blue Train was suspended during World War II to help with the war effort and was reinstated in 1946. It was in this post-war period that the train's local moniker, the Blue Train, named after the blue paint that decorated the steel carriages, became the official name of the luxury service. Since the mid-twentieth century, the Blue Train has upgraded its rolling stock multiple times, with air-conditioned dining and kitchen cars added to each train and sixteen new carriages added in 1972.

Travelling across a good portion of the country, the views of the South African landscape change as you pass through the semi-desert regions of the Great Karoo region, over the Great Escarpment of the South African Plateau, and into the coastal Cape Fold Mountains that surround Cape Town. The Blue Train offers a luxurious train experience that is hard to match.

The Dolbadarn was originally built in 1922 for the Dinorwic Railway. All of the rolling stock that was purchased by Lowry Porter had to be regauged to fit the new narrow gauge of the Llanberis Lake Railway.

THE LLANBERIS LAKE RAILWAY

Padarn Lake (Llyn Padarn in Welsh) is a glacially formed lake with a depth of ninety-four feet. The lake is also the longest lake in Wales.

The Elidir was originally built in 1889 for the Dinorwic Railway and is the oldest locomotive still operating on the Llanderis Lake Railway.

Another gem of a Welsh railway, the Llanberis Lake Railway is a heritage narrow-gauge railway that traverses the northern shore of Padarn Lake in Snowdonia National Park, Wales. The two and half mile route begins in the town of Llanberis and ends at the Pen Llyn terminus in the Padarn County Park. The trackbed of the one foot and eleven and a half inch gauge track runs along the old Padarn Railway that once connected the slate quarry of Dinorwic to Port Dinorwic on the Menai Strait. The Padarn Railway itself ran along the remnants of the even older Dinorwic Railway, which was completed in 1825. The Padarn Railway replaced the Dinorwic Railway in 1843 with horse drawn carriages and eventually implemented steam-powered engines 1848. This seven-mile long railway operated in conjunction with the quarry for over a century until they was closed in 1961.

The tracks to the Padarn Railway were removed from the trackbed in 1962, leaving little hope for another railway to take its place. But a train enthusiast from Southend-on-Sea, Lowry Porter, saw this empty trackbed as an opportunity to open a scenic railway along Padarn Lake. Porter proposed a shorter route along the trackbed that would travel along the northern shore of the lake, and in 1970 the County Council bought the trackbed to begin restoration. A very narrow gauge of track was installed, and Porter's budding railway company bought the quarry's rolling stock and land. By 1971, the Llanberis Lake Railway was opened to the public, operating with three steam locomotives as passenger trains and several diesel trains used as work trains and replacements for the steam engines when not in operation.

Right: *The old Dinorwic Quarry along the banks of Padarn Lake.*

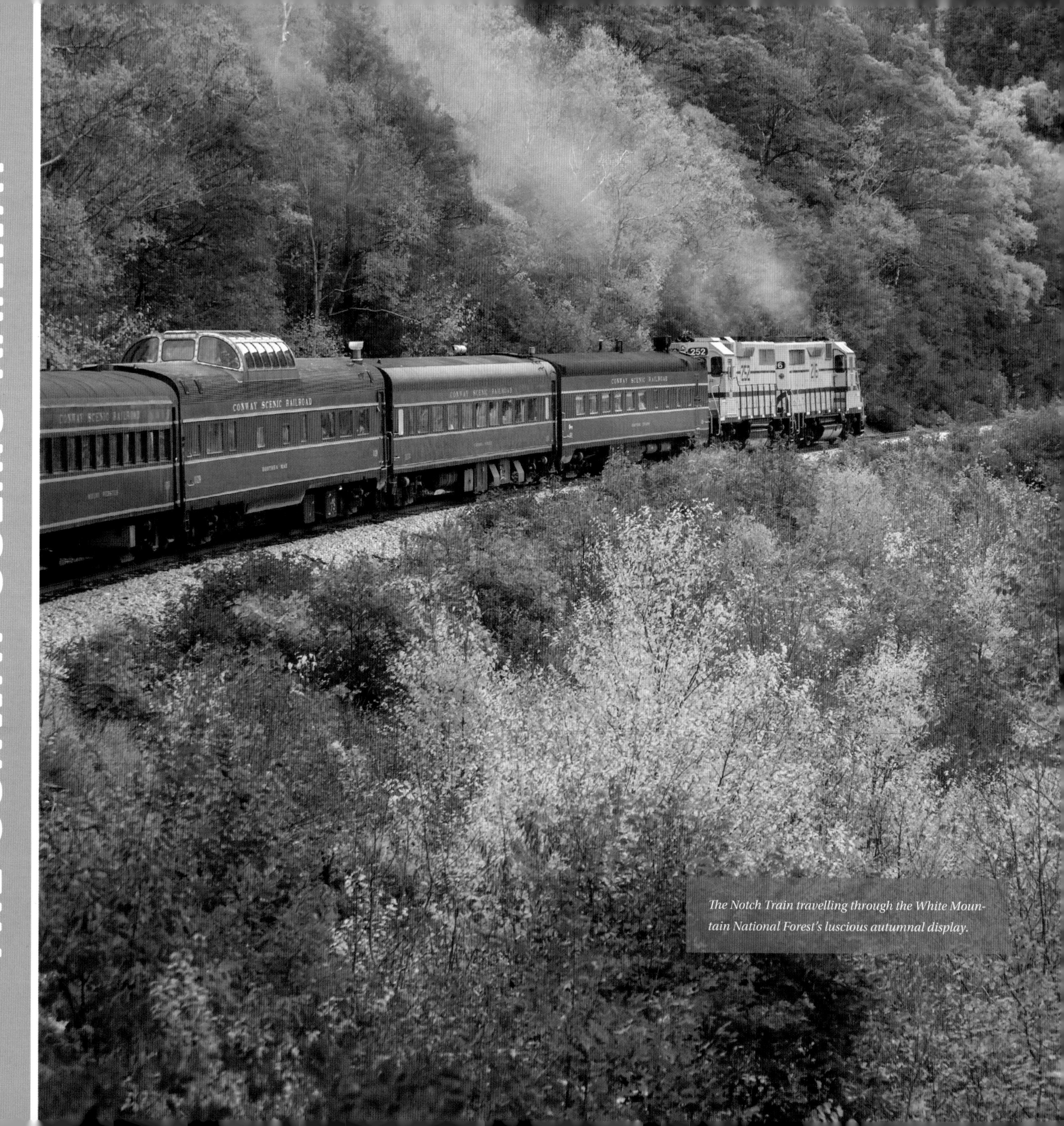

The Notch Train travelling through the White Mountain National Forest's luscious autumnal display.

Left: *The North Conway Depot and Railroad Yard was built in 1874, two years after the Conway Branch was completed.*

Below: *The Frankenstein Trestle along the Notch Train was built in 1875 and was originally supposed to be constructed out of wood. The schooner carrying the pine that was to be used to build the trestle was lost at sea. The iron that was subsequently used was a better choice in the long run.*

Bottom right: *The steel bridge called Fourth Iron crossing the Sawyer River. The Notch Train ascends nearly 1,700 feet in elevation in the thirty miles it travels between North Conway and Fayban.*

This scenic railway in New Hampshire operates two different routes on two formerly separate railways that had been abandoned for years, each departing from the town of North Conway either toward Fayban in the White Mountains or Conway in the Mount Washington Valley. The Conway Scenic Railway's "Valley Train" operates on the old Conway Branch of the Boston & Maine Railroad, built in 1841, from North Conway to Conway. These tracks have been owned and operated by various rail companies since the mid-nineteenth century. The infrastructure was installed by the Great Falls & Conway Railroad, which operated on the tracks until 1890 when the Boston & Maine Railroad acquired the rights to the tracks. The Boston & Maine Railroad used the tracks until 1974, two years before a local businessmen in the area bought the tracks and founded the first route of the Conway Scenic Railway.

The northern section of the Conway Scenic Railway, the "Notch Train," running through Crawford Notch State Park and White Mountain National Forest between North Conway and Fayban, was chartered in 1867 as the Portland & Ogdensburg Railway. Originally planned to connect Portland, Maine, with Ogdensburg, New York, the original route failed, and the tracks that were built between Maine and New Hampshire were leased as the Mountain Division of the Maine Central Railroad. Today, the "Notch Train" is a fifty-mile, five-hour round trip through the American Northeast's most luscious mountain forests.

Today, both routes operate out the of main terminal train station in North Conway, the North Conway Depot and Railroad Yard, which was listed on the National Register of Historic Places in 1979. This station originally operated as the northern terminus of the Conway Branch of the Boston & Maine Railroad. With Boston as the railway's southern terminus, the Conway Branch serviced the North Conway for eighty-seven years, giving travellers access to New Hampshire's summer resorts and winter skiing. The Conway Scenic Railway continues to give travellers access to some of New Hampshire's most stunning views with a luxury service that provides first class options, dome cars, dining cars, and even an open-air car to completely expose yourself to the fresh mountain air.

THE CALIFORNIA ZEPHYR

The California Zephyr travelling in front of the Book Cliffs between Green River and Floy, Utah.

Amtrak's second longest route in America, the California Zephyr provides service between Chicago and the San Francisco Bay area through some of the nation's most spectacular scenery. Before Amtrak took over the route, this epic passenger train was collectively operated by the Chicago, Burlington & Quincy, the Denver & Rio Grande Western, and the Western Pacific Railroads. Each claimed the route to be "the most talked about train in America." These operators wanted to provide passengers access to the Golden Gate International Exposition of 1939 and worked together to connect their railways together to create a route originally known as the Exposition Flyer. The Chicago, Burlington & Quincy Railroad provided service between Chicago and Denver, the Denver & Rio Grande Western Railroad provided tracks between Denver and Salt Lake City, and the Western Pacific Railroad ran from Salt Lake City to Oakland. This conglomeration operated for ten years, until they decided to streamline their service and rename it the California Zephyr in 1949.

Despite its popularity, the California Zephyr was not immune to the falling passenger rates that were striking the nation during the mid-twentieth century. The Denver & Rio Grande Western and the Western Pacific applied to discontinue their service in 1969, and the Western Pacific was permitted to do so providing that the Chicago, Burlington & Quincy and the Denver & Rio Grande Western continued to provide some semblance of the semi-transnational route. The last California Zephyr of the era ran on March 22, 1970.

The California Zephyr revived the next year as Amtrak consolidated its own rail network across the nation, but the Denver & Rio Grande Western refused to join Amtrak's rail system, leading Amtrak to re-route the Zephyr away from Colorado and onto the Union Pacific Overland Route through southern Wyoming. It wasn't until 1983, when Denver & Rio Grande Western decided to join Amtrak, that the California Zephyr's original track was reunited. The track no longer runs all the way to its Oakland terminus but now terminates at the train station in Emeryville, California.

Above: *A view of eastern Utah's high desert from the train.*

Below: *The vista-dome coach Silver Scout.*

Bottom: *Today, the California Zephyr operates with a baggage car, a transition sleeper, two sleeping cars, a dining car, three coaches, and a lounge car.*

The Swiss Alps contain the highest mountain tops of the Alps, including Dufourspitze, Dom, Liskamm, and the Matterhorn.

THE BERNINA EXPRESS

As the third highest railway in Switzerland, the Bernina Express offers views of the Swiss Alps that few other routes can provide along two historic railways, the Albula and Bernina Railways. Owned and operated by the largest privately-owned railway in Switzerland, the Rhaetian Railway, the Bernina Express connects the resort town of Chur, Switzerland, and Tirano, Italy, by way of the Bernina Pass (at 7,638 feet in elevation) in the Bernina Range of the Alps.

The Albula Railway, which also serves the Glacier Express, opened in 1904, as a steam engine, serving between Thusis and St. Moritz, Switzerland. It was later transitioned to an electric route in 1919 because of coal shortages during World War I. The Bernina Railway was built a few years later, obtaining concession to build in 1906 and not finished until 1910. The two railways are very much cherished for their influence on Italian/Swiss cross-border relations, and were listed as a UNESCO World Heritage Site in 2008 as the Rhaetian Railway in the Albula/Bernina Landscape.

The first years of operation were very financially tight for the independent Bernina Railway because of maintenance and construction costs needed to keep the railway safe for year-round service. The Rhaetian Railway acquired the route in the 1940s and has operated it since. As a part of the rail network of the Rhaetian Railway, the Bernina Express was born. Travelling through some of the most majestic and treacherous landscapes in Europe, the Bernina Express passes through fifty-five tunnels and crosses 196 bridges with ease during the four-hour trip.

Top right: *Amazingly, the Bernina Express operates year-round.*

Bottom right: *The Brusio Spiral Viaduct was necessary to be built so that the Bernina Express did not exceed its maximum gradient of seven percent. The viaduct was required south of the town of Brusio in order for the train not to slip on its way up or lose control on its way down.*

Left: *The Bernina Railway is one of the steepest adhesion railways in the world.*

THE GLACIER EXPRESS

The Landwasser Viaduct was finished in 1902 and was originally used for Rhaetian Railway. It is a significant landmark that is a part of the historic Albula Railway, which is a UNESCO World Heritage Site.

The Glacier Express connects the mountain resort towns of Zermatt and St. Moritz in southern Switzerland. Covering 181 miles, the Glacier Express crosses 291 bridges and passes through ninety-one tunnels—one of which, the Furka Tunnel, is nine miles long. There is really no better way to see the Swiss Alps. The famous Matterhorn is located outside the train's western terminus of Zermatt while the Albula Alps provide a backdrop to the eastern terminus of St. Moritz. The trip is a high elevation Alpine excursion ranging from nearly 2,000 feet above sea level at the lowest point near the town of Chur in the Grisonian Rhine Valley to almost 6,700 feet above sea level at Oberlap Pass.

Southern Switzerland is notorious for its scenic mountain landscapes and serene rock formations, but these characteristics made this area terribly inaccessible during winter months before train services came to the area. Many roads are closed during winter months, leaving train services, like the Glacier Express, as the only connection between these stowed away cantons. After the connection of two early twentieth century rail services in the area, the Visp–Zermatt Bahn (BVZ) and the Furka Oberalp Bahn (FO), the Glacier Express provided service for the first time between Zermatt and St. Moritz on June 25, 1930. In these early years of service, electric locomotives were used on the old BVZ while steam locomotives were used on the FO Route. The Glacier Express became an all-electric service when an electric overhead catenary was installed on the FO route in 1942. Year round operations were not available until 1982 when the Furka Base Tunnel was opened. Up until that point, the winters made the area impassable.

Today, the Glacier Express is a popular tourist destination that attracts nearly a quarter-million tourists every year. A trip from one end of the line to the other takes nearly eight hours, making it the "slowest express train in the world."

The Glacier Express passes through many mountainous hamlets that were once very isolated regions before rail service came to the area.

Panoramic cars allow passengers of the Glacier Express to see much of the beautiful landscape that surrounds them.

The Valle de Núria Rack Railway uses the Abt system for its rack infrastructure to transition from adhesion to rack traction.

THE VALLE DE NÚRIA RACK RAILWAY

Providing access to the isolated Valley of Núria, the Valle de Núria Rack Railway ascends into the Pyrenees Mountains from the Catalan municipality of Ribes de Freser through Querlabs, the highest point in the valley accessible by road, toward the shrine and resort of Núria, Spain. (The only other way to reach Núria is by footpaths or mule tracks through the mountains.) The Valle de Núria Rack Railway was opened in 1931 to provide a more convenient route into this secluded area, which is filled with local history.

Núria was a place of respite for Saint Giles, who fled from persecution by the Roman government in 700 AD. It was said that, having lived in the valley for four years, Giles left an image of the Virgin Mary in a cave, a cross, a pot used for cooking, and a bell to call shepherds to dinner in the valley. In 1072, a pilgrim named Amadéu traveled to the area and built a chapel for other pilgrims coming to area. He found the artifacts left behind by Saint Giles seven years later and placed them in the chapel. Today, the chapel and resort in Núria are what draw people to this area.

The Valle de Núria Rack Railway operates under conventional rail adhesion for the first three miles of the trip before it is converts into a rack railway. Over the near eight-mile length of the track, the railway overcomes 3,484 feet in elevation at a fifteen percent gradient.

***Above:** Núria was the drafting place of Catalonia's first Statute of Autonomy, which concedes self-government to the region at a subnational level.*

***Left:** The train station at Núria is attached to the resort to keep passengers away from the brutal winter weather during skiing season.*

THE CHIHUAHUA-PACIFIC RAILWAY

The Chihuahua-Pacific Railway climbing out of Copper Canyon.

Service between Témoris, Chihuahua, and Los Mochis, Sinaloa, on the Chihuahua-Pacific Railway was temporarily shut down in the fall of 2018 because the tracks were damaged from flooding during a tropical storm.

The Chihuahua-Pacific Railway is also known as El Chepe.

Conceived in 1880 and completed in 1961, there is perhaps no other railway in the world that has taken so long to be constructed as the Chihuahua-Pacific Railway. In 1880, the president of Mexico at that time, General Manuel González, realized the need for a railway to traverse the Sierra Madre Occidental range to connect the port of Topolobampo near Los Mochis, Sinaloa, to Chihuahua, Chihuahua. A concession was granted to Albert Kinsey Owen, who was a member of the Utopian Socialist Colony of New Harmony, Indiana. Owen's ultimate plan was to settle a new socialist community whose settlers would also serve as the development company for the railroad.

Owen dreamt of a transcontinental railroad from Galveston, Texas, to Topolobampo, Sinaloa, but American investors were not convinced of his scheme. The railroad that Owen had fought for was slowly losing steam, and those who moved to Mexico under Owen's convincing ideals began to move back to where they came from. Owen gave up on his socialist paradise, and construction of the railway was passed along to another hopeful railway magnate of the Kansas City, Mexico & Orient Railway, Arthur Edward Stidwell. Stidwell's company began construction in 1900. But due to financial difficulties and dangerous terrain, the whole railway was not completed until 1961. The Mexican government owned the railway from its opening in 1961 until 1998, when it sold the railway to Ferromex, the largest privately owned railway company in Mexico.

Today, the Chihuahua-Pacific Railway offers passengers first- and second-class travelling options. There are two trains that run a day: an express train and a slower train that stops at fifteen official stations while also allowing for flag stops where needed. The train travels 418 miles over thirty-seven bridges and through eighty-six tunnels across the Sierra Madre Occidental range. The train reaches massive heights above sea-level, nearly 7,900 feet, at the Divisadero, or the Continental Divide. A trip in one direction takes sixteen hours, but the sights of the mountains, desert, and Copper Canyon will keep you more than occupied.

A train station in the state of Chihuahua. Although the state of Chihuahua is associated most with its namesake, the Chihuahua Desert, the area has more forests than any other part of Mexico aside from the Durango area.

The section of tracks built between Voss and Oslo were standard gauge, which required the narrow-gauge tracks that were used between Bergen and Voss to be replaced before trains could complete a full trip between Bergen and Oslo.

THE BERGEN RAILWAY

There are a total of thirty-nine stations along the Bergen Railway. The railway can reach a maximum of ninety-nine miles per hour.

The Bergen Railway originally operated with steam engines until the railway was completely electrified in 1964.

The Bergen Railway is an example of fortitude and resilience. Building a railway between Norway's two largest cities, Oslo and Bergen, is no easy task, especially in a country where winter temperatures often range well below zero. Han Gløersen, a writer for a local newspaper, conceived the railway in an article he wrote on August 24, 1871, proposing a connection between Norway's largest economic and cultural hubs. But the railway that was first built leading out of Bergen, on Norway's west coast, only made its way sixty-eight miles east to Voss at first.

Construction for the Voss Line began December of 1875, and the line was opened eight years later in 1883. After its completion, Norway fell into a recession that nearly put a halt to all public works projects in the nation, but in 1894, after fives days of debate, Norway's parliament decided that it would be best to continue building the connection to Oslo. Financing for the line was ready within the year, but it took another six years to survey the railway's mountainous route. Roads being built along the route to transport supplies were finished by 1902, and construction on the railway finally began that year. High-altitude conditions, several feet of snow, and below-zero temperatures made construction extremely difficult. This section of the railway has 113 tunnels along its route, totaling to nearly seventeen miles of tunnel. The longest tunnel along the route, also the longest tunnel north of the Alps, the Gravehalsen Tunnel, took six years to complete and is just over three miles long. The complete Bergen Railway officially opened on December 9, 1907, but the train had to return because it got stuck in the snow—it then took another month and a half to clear the tracks of snow. The first train to successfully leave from Oslo and make it to Bergen departed on June 10, 1908.

The Bergen Railway is used for express trains from Bergen to Oslo, freight trains, and commuter trains from Mydral to Bergen. Along the total length of the 306 miles between Oslo and Bergen, there are 182 tunnels, totaling to forty-five miles in length. The Bergen Railway offers spectacular views of Norway's glacially-formed geography, mountainous terrain, deep fjords, and the epic Hardangervidda Plateau.

A surreal photo of a steam engine pushing a rotary snowplow through the snow in 1908.

THE BELMOND ANDEAN EXPLORER

The Belmond Andean Explorer travelling through the Andes.

The Belmond Andean Explorer provides a unique experience as South America's first luxury sleeper train. Departing from Cusco, Peru, the train climbs the Altiplano, or the Andean Plateau, which is the widest section of the Andes and the second largest area of high-elevation plateau in the world outside of Tibet, toward Puno, Peru. In Puno, the train stables for the night along the shores of Lake Titicaca, the largest lake in South America.

During your stay in Puno, you can take boat tours on the lake to any number of its many islands in the morning before you return to the train to continue your journey towards Arequipa City. Along the route to Arequipa City you will descend from the Altiplano and stop to visit the Sumbay Caves, which have petroglyphs believed to be nearly 8,000 years old. At this stop, you can also take a trip to take in the views of Colca Canyon, a very popular tourist destination in the country receiving nearly 120,000 visitors a year. After a brief stop at the Sumbay Caves and Colca Canyon, the Andean Explorer continues to its terminus of Arequipa City.

The Belmond Andean Explorer's train set was originally used on the Great South Pacific Express in Australia between 1999 and 2003 before it was shipped to Peru in 2016. There are various accommodation levels that the Belmond Andean Explorer provides, with bunk- to double-bed en-suite cabins, two dining and bar cars, and an open-air observation car.

Top left: *During your stay in Puno, boat tours on Lake Titicaca will take you to the Uros Islands, which are man-made islands composed of reeds that are weaved by the Uru people.*

Top right: *The views from the top of Colca Canyon are astonishing. The canyon is one of the deepest canyons in the world, with a depth of nearly 10,000 feet.*

Right: *Each luxury trip can hold up to forty-eight passengers.*

The Bavarian Zugspitze Railway was nominated for a Historic Landmarks of Civil Engineering in Germany Award in 2007.

THE BAVARIAN ZUGSPITZE RAILWAY

Left: Descending from Zugspitzplatt Station, you are presented with views of Lake Eibsee.

Below: The Bavarian Zugspitze Railway in the Rosi Tunnel.

One of four remaining rack railways still in operation in Germany, the Bavarian Zugspitze Railway travels from the resort town of Garmisch-Partenkirchen to the base of the Zugspitze, which is the tallest mountain in Germany. The railway opened in three stages. The first stage was a two mile section of track that opened on February 19,1929, between Grainau and Eibsee, Germany. The second stage connected Grainau to the starting point of the railway in Garmisch-Partenkirchen on December 19, 1929, with a standard adhesion track that runs for nearly five miles. The last stage from Eibsee to Schneefernerhaus was finished July 8, 1930, introducing the last five miles of track to the summit of Zugspitze.

The railway operates both as a standard adhesion railway and a rack railway, using the rack system to ascend nearly 6,000 feet at a twenty-five degree gradient. Since 1992, the Schneefernerhaus terminus, found at the end of the two-mile long Zugspitze Tunnel, was closed, and service was then only provided to the Zugspitzplatt Station, which opened in 1987. The Zugsitzplatt Station sits on a plateau at the base of Zugspitze as opposed to the Schneefernerhaus Station, which climbed to the mountain's summit. The Zugspitzplatt Station uses the newly constructed Rosi Tunnel to make it to its terminal point, using about three-quarters of the Zugspitze Tunnel before it branches off toward a lower point along the mountain.

The Bavarian Zugspitze Railway is a relatively short railway, but the height in which it ascends is no small feat. It has the greatest height difference of any other railway in Europe. It is also the third highest railway in Europe and the highest railway in Germany.

The train pictured in 1931.

THE STOURBRIDGE LION

Throughout the late nineteenth century, railroads both knowingly and unwittingly created a mythology of early railroading. There were no "first railroad cars" like this in the United States, but the image provided a nice contrast with luxury trains being advertised to the public after the Civil War.

Only at the beginning did British technology play a substantial role in American railroading. The first locomotive in the U.S. was an import from the machine works of John Rastrick of Stourbridge, England, which delivered the "Stourbridge Lion" to the Delaware & Hudson Canal Company. After a nearly disastrous trial on the company's eighteen-mile tramroad in eastern Pennsylvania in August of 1829, the "Lion" and its sister engine were set aside. Later British locomotives, such as the "Planet" type locomotives built by Robert Stephenson, were far more successful. In 1831, the Camden & Amboy imported the "John Bull," now preserved at the Smithsonian Institution.

A replica of the "Stourbridge Lion," the first steam locomotive used in America, built in 1828 for the Delaware & Hudson Canal Company.

A lithographic print of the Stourbridge Lion. The Stourbridge Lion's first run was on August 8, 1829. It was retired five years later in 1834 and is currently owned by the Smithsonian Institution.

***Right:** Arriving here from England in 1831, the "John Bull" was modified substantially to make it better suited to American conditions.*

The Algoma Central Railway provided year-round access to roadless areas in the Agawa Canyon three days a week.

THE ALGOMA CENTRAL RAILWAY

Travelling between the border city of Sault Ste. Marie, Ontario, Canada, and the town of Hearst, Ontario, Canada, the Algoma Central Railway was chartered in 1899 to Francis H. Clergue, a high-level American industrialist in Sault Ste. Marie, who wanted to transport resources such as timber and iron ore from northern Ontario to his processing facilities along the Great Lakes. The charter originally allowed for Clergue to build a branch north from Sault Ste. Marie toward a junction with the Canadian Pacific Railway, and another branch that would run along the coast of Lake Superior to a harbor outside of Wawa, Ontario.

Clergue was expanding his infrastructure quickly and soon gained another charter for the Ontario, Hudson Bay & Western Railway which was planning on building a connection between the Canadian Pacific Railway and the Hudson Bay. With plans of creating an adequate shipping network in the isolated area, Clergue renamed his railroads the Algoma Central & Hudson Bay Railway Company. But too rapid of an expansion led to the bankruptcy of the railway in 1903, and all construction on the line stopped. The railway only made it fifty-six miles north of Sault St. Marie, with no connection made to other railways. Construction resumed a few years later in 1909, finally allowing the train to connect with Michipicoten Harbour at Hawk Junction, the Canadian Pacific Railway with a junction at Franz, and the Canadian Northern Railway with a junction at Oba. The Algoma Central & Hudson Bay Railway made it to Hearst in 1914, which would become its ultimate terminal point some 150 miles away from James Bay.

By the 1960s, the Algoma Central Railway had dropped the "Hudson Bay" part of its name and began offering tourist services along its railway through the Agawa Canyon. The Agawa Canyon is 114 rail miles north of Sault Ste. Marie and was carved by the Agawa River. The Algoma Central Railway follows the Agawa River through the canyon's isolated natural beauty. The railway was bought in 1995 by the Wisconsin Central, which itself was acquired by Canadian National in 2001.

In 2014, Canadian National announced that they would discontinue the passenger service because they could no longer sustain their service without the help of government subsidies—which were being cut by the Canadian government. Railmark Canada Ltd. announced it would take over the passenger service, and the Canadian government assured another three years of funding. But Railmark failed to find a line of credit to operate the train, so the fate of the service is not guaranteed after the subsidies are no longer available.

Top right: *The Algoma Central Railway ran some of the longest passenger trains in North America with up to twenty-four cars on one trip.*

Bottom right: *In 2014, the annual average ridership was estimated to be just over 10,000 passengers.*

THE GONO LINE

The Resort Shirakami train on the Gono Line.

Operated by the East Japan Railway Company between the Higashi-Noshiro Station in Akita Prefecture and the Kawabe Station in Aomori Prefecture in the north of Japan's main island, Honshu, the Gono Line is a scenic route that travels along the Sea of Japan's coastline. There are forty-three stations along the ninety-one miles of non-electrified track. The original Gono Line, opened in 1909, was operated by Japanese Government Railways and only connected between the Higashi-Noshiro and Noshiro Stations—an extremely short distance for a train. The line was expanded to Iwadate in 1926 and then Mutsu-Iwasaki in 1932, totaling nearly thirty-two miles of track.

The line reached its full length when the privately held Mutsu Railway connected Kawabe Station to Mutsu-Akaishi Station in 1936. The Mutsu Railway and the Gono Line were renamed to the Gono Line that same year. Since then, the route has operated as a commuter train and tourist attraction to provide sights of the Sea of Japan. All trains, even the express trains, must stop at all of the designated stations, except for the Resort Shirakami service. The Resort Shirakami operates as a scenic passenger train along the Gono and Ou Main Lines. It provides a five-hour trip along the coast of northern Honshu that has been in operation since 1997.

Right: *All of the trains that operate on the Gono Line are diesel multiple unit engines, which provide power with on-board diesel engines that do not require the use of an external locomotive to transport carriages.*

Bottom left: *The Sea of Japan provides a beautiful backdrop outside the windows.*

Bottom right: *The Sea of Japan has a near nonexistent tide due to its enclosure from the Pacific Ocean, causing a serene seascape to gaze upon.*

THE SAGANO SCENIC RAILWAY

Left: *The Hozu River is a scenic river that starts in the mountains near Kameoka and then snakes down the mountains toward the Arashiyama section of Kyoto.*

Below: *Passengers are able to take in the mountain air with the open carriages of the Sagano Scenic Railway.*

Bottom right: *The Japanese Maples turning colors in the fall make this scenic railway even more scenic.*

Operated by the West Japan Railway Company (JR West), the Sagano Scenic Railway travels for four and a half miles on the superseded tracks of the original Sagano Line, which is a part of the San'in Main Line. Today, the Sagano Scenic Railway operates along four stops in suburban Kyoto, from the Torokko Saga Station in Arashiyama through the Hozukyo Ravine to the Torokko Kameoka Station in Kameoka. The original Sagano Line was a part of Kyoto Railway and opened in 1899 to serve the northern parts of Kyoto Prefecture.

In 1907, the railway was nationalized and became a part of the San'in Main Line, which served a larger area from Kyoto to Shimonoseki in Yamaguchi Perfecture. Traveling through the mountains outside of Kyoto, the Sagano Line was built on relatively level ground that bypasses the steep gradients of the surrounding mountains. In 1989, a shorter, faster, and straighter bypass between the Saga-Arashiyama and Umahori Train Stations was constructed, abandoning the original route. The abandoned route reopened in 1990 as the Sagano Scenic Railway, focused on being a tourist destination for passengers to enjoy the scenery of the Hozu River Gorge.

The Sagano Scenic Railway does not operate during the winter or on Wednesdays. With five open carriages led by a diesel locomotive, the Sagano Scenic Railway meanders through copses of Japanese maples and cherry blossoms that provide beautiful foliage in the spring and fall. All seats are reserved, so booking ahead of time is required to enjoy this route.

The railway originally opened in sections, with the section between Gifu and Kagamigahara, Japan, opening in 1920, and the rest of the line between Kagamigahara and Toyama in 1934.

THE TAKAYAMA MAIN LINE

The Takayama Main Line operates for 140 miles between the cities of Gifu and Toyama, Japan, in their respective prefectures. Also known as "The Hida" or the "Wide View Hida," the Takayama Main Line has gained popularity for the scenic views it provides of the historic and mountainous Hida Province in northern Gifu Prefecture. Gifu is part of the larger metropolitan district of Nagoya, and it provides many tourist attractions such as the Nagoya Castle and the Gifu Great Buddha Statue in Shohoji Temple. From the city of Gifu, the Takayama Main Line travels northwest through forty-five stations to the city of Toyama in the Hokuriku district, a culturally traditional district that receives a high volume of snowfall unmatched by any other inhabited and arable region in the world.

Once used as a major mode of transportation between Gifu and Toyama Prefectures, it is mostly ridden today as a scenic railway to take in the sights of the Hida Mountains, a mountain range that reaches nearly 10,000 feet above sea level with numerous peaks that are as high as 9,000 feet above sea level. Today the Tokaido Shinkansen high-speed rail and the Hokuriku Main Line offer faster ways to travel between prefectures. As you travel along the Takayama Main Line, you will pass many natural hot springs, the Kiso River, historic towns like Shirakawa and Gokayama, and a portion of the Japanese Alps in the Hida Mountains.

The Takayama Main Line uses diesel locomotives and is operated by various railway companies. The Central Japan Railway Company provides the service between the Gifu and Inotani Train Stations, while the West Japan Railway Company provides service between Inotani and Toyama Train Stations—both of which provide train service to hundreds of millions of passengers a year. The Takayama Main Line is covered by the Japan Rail Pass, so there are no special hoops you need to jump through to get a seat on this train.

Top right: *An autumnal view of the Takayama Main Line in the Hida Mountains.*

Bottom left: *A Takayama Main Line train outside the city of Gero in Gifu Prefecture.*

THE KUROBE GORGE RAILWAY

The name Kurobe has no meaning in Japanese, but the Ainu translation of the word means "shady river." Ainu is a language spoken by the Ainu ethnic group from the northern Japanese island of Hokkaido. Ainu has no genealogical relationship to any other language family.

Left: *The Kurobe Gorge Railway crossing the Shin-Yanabiko Bridge.*

Top right: *Passengers in an open carriage, exposed to the cool mountain climate.*

Bottom right: *The Korube Dam is the tallest dam in Japan at 610 feet high. It was built between 1956 and 1963 for a total cost of ¥51.3 billion.*

The Kurobe Gorge Railway was built by the Kansai Electric Power Company to help transport construction supplies to its power stations in the Kurobe Gorge, inclduing the Kurobe Dam, Japan's tallest dam built to help meet energy needs during the rapid development of post-war Japan. The Kurobe Gorge Railway is a privately-held railway that follows the Kurobe River through its gorge in the Hida Mountains of Toyama Prefecture. Traveling from the town of Unazuki to the terminal point of Keyakidaira Train Station, the passenger service of the Kurobe Gorge Railway was opened in 1953 during the construction of the Kurobe Dam, which was finished in 1963. The original section of track travelled from Unazaki to Nekomata in 1923, and was then electrified and expanded toward its current terminus of Keyakidaira in 1937. The Kurobe Gorge Railway does not travel all the way to the Kurobe Dam though. From the terminus of the Keyakidaira Train Station, the Kurobe Senyo Railway, an underground railway that is not open to the public, completes the journey to the dam through rugged terrain.

The recreation and the scenery offered along the Kurobe Gorge Railway is unlike any other railway in the world. The Kurobe Gorge has long been seen as a valuable natural resource to the Japanese people. The powerful forces of the Kurobe River have carved the gorge the railway travels through. The river begins at the top of Mount Washiba and then rapidly descends the Hida Mountains toward the Sea of Japan. Its steep descent makes it a great source for hydroelectric power. Although the train does not travel all the way to the Kurobe Dam, you can see two other dams along the river as you pass through the gorge. There are also a number of hot springs along the river that passengers can enjoy as they explore the surroundings of each stop.

The Kurobe Gorge Railway has been owned by the company Kurotetsu since 1971, after the Kansai Electric Power Company created it as a subsidiary to operate and manage the railway. It operates with 27 electric locomotives, 138 passenger carriages, and 322 freight wagons. You can pay for a variety of service tiers that will provide you with degrees of more or less comfort, as well as more or less protection from the changeable mountain climate.

The Waumbek locomotive operating on Mount Washington. Built in 1908, this locomotive was converted for a short time to burn biodiesel but was reverted back to coal.

THE MOUNT WASHINGTON COG RAILWAY

The first mountain-climbing and second steepest cog railway in the world, the Mount Washington Cog Railway of New Hampshire ascends the highest mountain in the Northeastern United States, Mount Washington. The railway may sound impressive by these stats, but the actual train ride might not be as exciting as you'd expect. Travelling at an average ascent speed of three miles per hour, it takes the Mount Washington Cog Railway sixty-five minutes to go the three miles up the mountain's western slope. The train travels at an average speed of five miles per hour on its descent, taking another forty minutes to complete this slow and arduous trip.

Conceived of and built by Sylvester Marsh in the mid-1800s, the idea of the Mount Washington Cog Railway was often thought of as an impossible dream, but Sylvester Marsh thought otherwise. Marsh was given the charter to build the railway, despite the harm and risk of injury it presented to its future ridership, because the state legislature never thought that such a railway could ever be built. By 1858, Marsh attained the charter to build the road that would bring construction supplies to the track, but the plan was put on hold as the Civil War commenced. It wasn't until 1866 that construction began. Marsh built a working prototype and track of his cog railway—the first successful cog railway in America—and began to attract investors. Passengers started paying for rides in August of 1868, although the railway was not completely finished until July 1869. Even the president of the United States at the time, Ulysses S. Grant, took a ride to the top of Mount Washington in August of 1869.

There are only two stations along the Mount Washington Cog Railway, one at the base of Mount Washington and one at the summit—three miles apart. The train gains over 3,000 feet in elevation, starting at 2,700 feet above sea level and reaching 6,288 feet above sea level with an average gradient of twenty-five percent. The maximum gradient the train operates at is a jaw-dropping thirty-seven percent on the cog system that Marsh himself invented and patented. The railway operated strictly with steam engines until 2008, when faced with rising criticism for environmental concerns and falling ridership. That year they introduced the first diesel locomotives on the railway in order to reduce reliance on steam engines and coal.

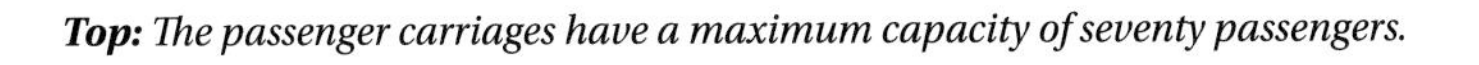

Top: *The passenger carriages have a maximum capacity of seventy passengers.*

Center: *The environmental costs of the railway are very real. Each three-mile run of a steam locomotive on this railway requires 2,000 pounds of coal and 1,000 gallons of water.*

Bottom: *The train passing the cairn that memorializes the death of Lizzie Bourne, who died from cold-exposure and exhaustion while climbing Mount Washington in 1855.*

THE STRASBURG RAILWAY

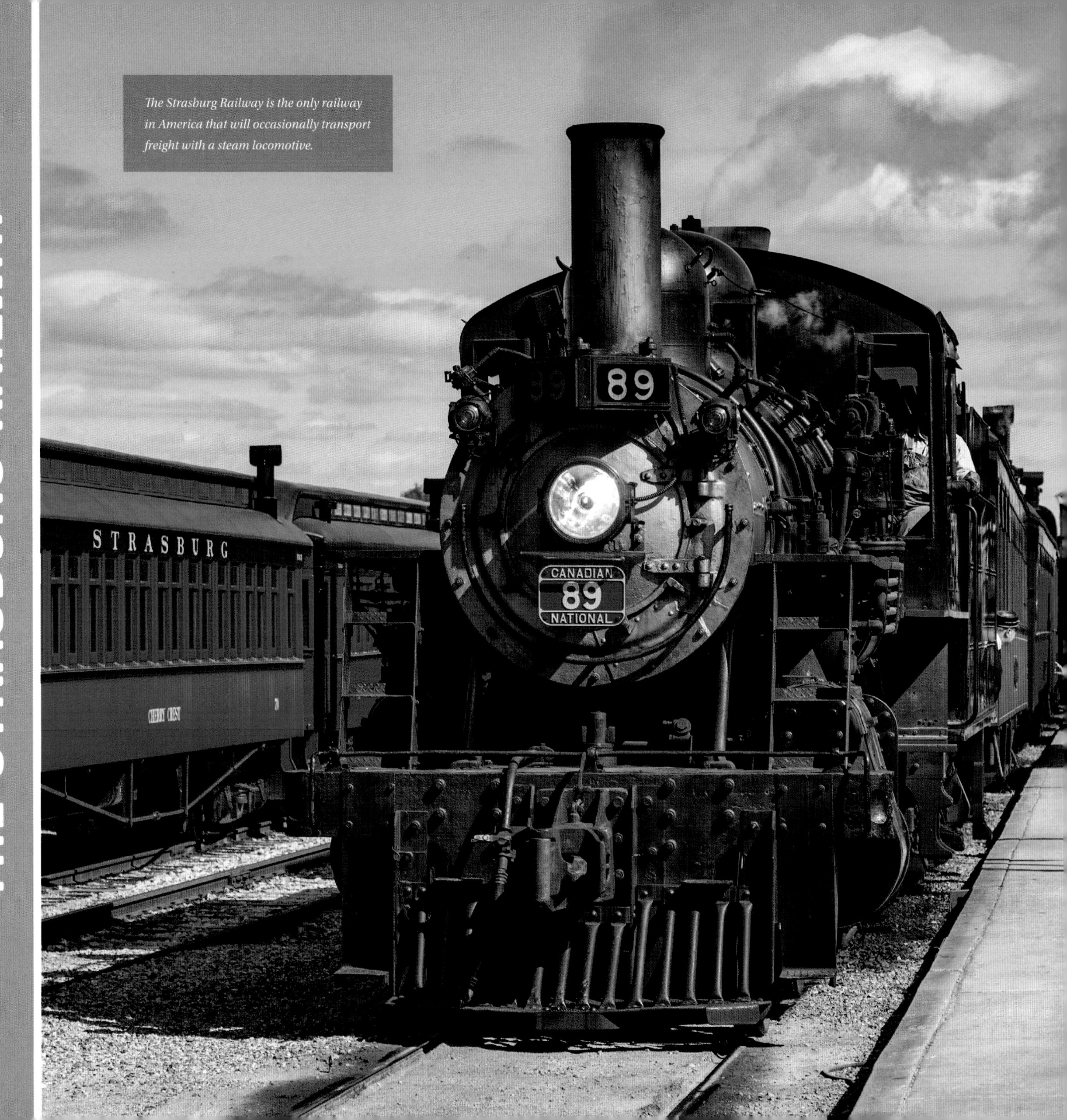

The Strasburg Railway is the only railway in America that will occasionally transport freight with a steam locomotive.

The Strasburg Railway was completed in 1837 as a freight service between Strasburg, Pennsylvania, and the junction at Leaman Place in order to save the economic future of Strasburg during the mid-nineteenth century's changing transportation mileu. In the early nineteenth century, transporting goods by canal was becoming more and more popular, leaving wagon roads and the Conestoga Wagon in the economic dust. With the opening of the Susquehanna Canal in southern Pennsylvania and the construction of the Philadelphia & Columbia Railroad, the goods that once passed through Strasburg on a wagon road to Philadelphia were no longer passing through town. Foreseeing a bleak outcome for Strasburg once its wagon route was obsolete, local businessman petitioned the state government to build a railway that would connect Strasburg to the Philadelphia & Columbia Railroad. The governor granted the charter for the railroad on June 9, 1832.

The early history of the railroad is not well known, but the grading of the roadbed began in 1835, with the tracks in operation two years later. The railroad operated with horse-drawn carriages until it was able to afford its first steam-engine locomotive. Throughout the nineteenth century, various owners controlled the operations of the railroad, passing from hand to hand. The railroad went through economic depressions, fire damage, and near abandonment, but has survived to this day due to the passion of local enthusiasts. Declining freight business pushed the estate of former Pennsylvania State Senator John Homsher, who took control of the railway in 1918, to file a public utility abandonment with the state commission in 1958. But a group of local enthusiasts were able to raise enough funds to buy the railroad and keep it in operation as a heritage line running through Pennsylvania's Dutch Country.

Built in 1910, the Strasburg Railroad locomotive #89 was originally used on the Grand Trunk Railway that traveled from Ontario, Canada, into the Northeastern United States.

Built in 1924, the Strasburg Railroad locomotive #90 is the railroad's largest and strongest engine. It was originally used for the Great Western Railway of Colorado.

Tourist excursions began on January 4, 1959, and have continued to this day. The Strasburg Railway is the oldest continuously running railroad in the Western Hemisphere and also the oldest public utility in the state of Pennsylvania. For nearly five miles of historic Pennsylvania Dutch Country, passengers can travel on one of five historic steam engines along with the railway's large fleet of wooden passenger coaches. Today the railway no longer connects with the Philadelphia & Columbia Railway but Amtrak's Philadelphia to Harrisburg Main Line.

THE GREAT SMOKY MOUNTAINS RAILROAD

Operating on fifty-three miles of the old Murphy Branch of western North Carolina, the Great Smoky Mountains Railroad travels through the scenic Nantahala National Forest between the mountain towns of Dillsboro and Andrews, North Carolina. Travelling through "fertile valleys, a tunnel, and across river gorges," according to the Great Smoky Mountains Railroad, the tourist train crosses some of Appalachia's finest terrain just south of the Great Smoky Mountains National Park.

The tracks were originally built as the Murphy Branch on what was then called the Western North Carolina Railroad. The Murphy Branch opened the mountainous and isolated areas west of Asheville, North Carolina, to the larger circles of commerce and culture that were connected to the Western North Carolina Railroad. Built between 1881 and 1884 with convict labor, the railroad shipped timber and other products from the mountains for nearly a century until the tracks laying west of Sylva, North Carolina, were closed due to declining freight traffic. The North Carolina Department of Transportation bought the portion of tracks that ran west from Sylva in 1988 and granted the right of way between Dillsboro and Andrews to the Great Smoky Mountains Railroad. In 1996, the Great Smoky Mountains Railroad bought the remaining section of track from the NCDOT.

Today the Great Smoky Mountains Railroad runs nearly 1,000 excursions a year on the tracks, travelling through the Great Smoky and Blue Ridge Mountains. You'll see over 125 varieties of trees, including hemlock, mountain ash, sugar maple, and umbrella magnolia, as well as the opportunity to see more than 100 types of fish and animals. This section of Appalachia is filled with waterfalls, rushing rivers, and picturesque streams that will captivate your eye and imagination. The railroad no longer operates all the way to Andrews due to damaged tracks. Service now runs for thirty-eight miles between Dillsboro and Nantahala.

Right: *Steam locomotive #1702 was built in 1942 and was acquired by the Great Smoky Mountains Railroad in 1994. It was taken out of service in 2004, underwent a complete restoration, and returned to service in 2016.*

Bottom left: *The railroad owns seven diesel locomotives, two of which are no longer operational because they were wrecked during the filming of the movie The Fugitive.*

Bottom right: *Gradients on the Great Smoky Mountains Railroad reach four percent in two areas.*

The Cumbres & Toltec Scenic Railroad operates one train a day from both the Chama and Antonito terminuses.

THE CUMBRES & TOLTEC SCENIC RAILROAD

A narrow-gauge heritage railway travelling between Antonito, Colorado, and Chamas, New Mexico, the Cumbres & Toltec Scenic Railroad gets its name from two of the major geographic landmarks it passes en route. On its sixty-four mile journey, it travels over the 10,000 foot Cumbres Pass of the San Juan Mountains and the Toltec Gorge of the Rio de los Pinos in the Rio Grande National Forest. The Denver & Rio Grande Western Railroad built the track in 1881 as part of the Alamosa-Durango Line that ran between Alamosa and Durango, Colorado. The Denver & Rio Grande Western Railroad was a pioneer in American mountain locomotion, travelling through the Rockies to connect Denver with Salt Lake City, Utah. It was instrumental in transporting the rich mineral resources of Colorado to the rest of the nation.

The Cumbres & Toltec Scenic Railroad is one of the few portions of track still used today that was part of the Denver & Rio Grande Western Railroad's San Juan Extension. Only the Durango & Silverton Narrow Gauge Railroad, the standard gauge San Luis & Rio Grande Railroad running out of Alamosa, Colorado, and the narrow-gauge Cumbres & Toltec Scenic Railroad remain. The Denver & Rio Grande Western Railroad began to abandon the Cumbres & Toltec Scenic Railroad in 1968 due to declining freight operations, but rail enthusiasts prompted the states of New Mexico and Colorado to buy a sixty-four mile section of the track for preservation. Both states are now part owners of the track and have developed the Cumbres & Toltec Scenic Railroad Commission to operate the railroad as a passenger excursion service. The Friends of the Cumbres & Toltec Scenic Railroad was established in 1988 to help create educational programs about the railroad's history.

The Cumbres & Toltec Scenic Railroad operates a handful of historic steam locomotives with a mixture of flat-roofed and clerestory carriages. There are three tiers of service on the route, including Coach, Tourist, and Parlor services. The railroad was listed on the National Register of Historic Places in 1973, and the boundaries of that designation were expanded in 2007. In 2012, the railroad was officially listed as a National Historic Landmark. In recent years, service has been cancelled for short amounts of time due to roadbed damage and threats from forest fires. In 2010, the Lobato Trestle, just east of Chama, was severely damaged by fire, causing a shortened route for trains travelling from each direction. The trestle was repaired in 2011 and is back in use.

Above: *The Cumbres & Toltec Scenic Railroad has been featured in numerous movies and documentaries, including the classic Indiana Jones and the Last Crusade.*

Below: *Heading east from Chama, New Mexico, and beginning its ascent to Cumbres Pass.*

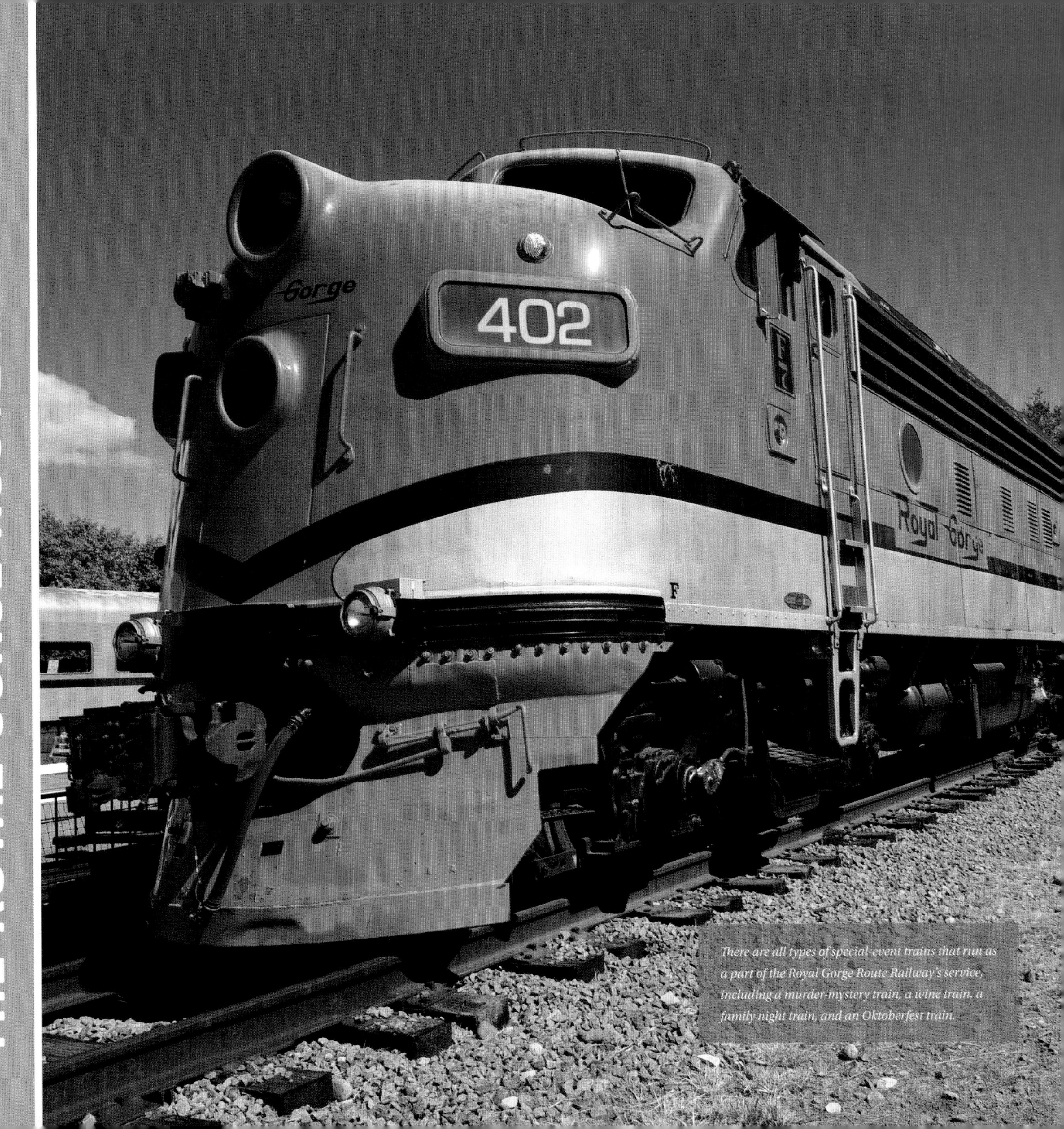

There are all types of special-event trains that run as a part of the Royal Gorge Route Railway's service, including a murder-mystery train, a wine train, a family night train, and an Oktoberfest train.

The Royal Gorge is six miles long, walled with cliff faces of dense igneous rock, and can reach depths of 1,000 feet. It was almost an impossible barrier to overcome when constructing the railway.

The Arkansas River is the sixth-longest river in the U.S., and second longest river in the Mississippi-Missouri river system. It originates in the Rockies just outside of Leadville, Colorado.

Also a vestige of the famed Denver & Rio Grande Western Railroad, the Royal Gorge Route Railroad of Colorado passes through what some may say is the most exquisite route the former Rocky Mountain railroad company built. Once a point of contention for competing railway companies, the Denver & Rio Grande Western and Santa Fe Railroads, to build a route through, the Royal Gorge of the upper Arkansas River only had room for one. In the late 1870s, miners flocked to the Arkansas River valley in Colorado in hopes of finding carbonate-rich lead and silver ore, leading the railway companies to follow.

With the influx of mining operations in the Arkansas River valley, both the Sante Fe and Denver & Rio Grande Western Railroads looked to build a line through the narrow pass at Royal Gorge in order to reach Leadville. After taking their dispute to the United States Circuit Court, the Santa Fe Railway was given the right to construct the railway in the gorge, while the Denver & Rio Grande Western was given secondary rights to construct their own tracks as long as they did not interfere with Santa Fe's tracks. The Denver & Rio Grande Western began laying track west of the gorge as Santa Fe began working its tracks through the gorge. Both crews were victims of sabotage committed by the competition. And Santa Fe even hired Doc Holliday and other hired guns to take over train stations along the Denver & Rio Grande Western route. Rights to the land swayed back and forth between companies, and the legal battle went on for years, but the dispute was eventually resolved out of court. Santa Fe conceded the rights of use to Denver & Rio Grande Western and received $1.8 million in payment for the track it built in the gorge. By 1880, the tracks made it up the friendly one percent water-grade ascent through the gorge to Leadville. Passenger service began that year and continued until 1967. Freight continued to be hauled on the line until 1989.

Sold off and merged a handful of times since the days of Denver & Rio Grande Western ownership, a twelve-mile section of the gorge railway was bought from Union Pacific by the newly incorporated Royal Gorge Express in 1997. Trains that pass through the gorge depart year-round from Cañon City, Colorado, toward its terminus of Parkdale, Colorado.

THE PILATUS RAILWAY

The Pilatus Railway with Lake Lucerne in the background.

The steepest mountain railway in the world, with a maximum gradient of forty-eight percent, the Pilatus Railway travels between Alpnach, Switzerland, to near the summit of Esel Mountain. The railway was built and designed by engineer Edward Locher, who raised and spent his own money to build the railway without the help of the government. Originally proposed in 1873 as a standard-gauge track with a maximum twenty-five percent gradient, the original route was much too costly to be built. The plan proposed by Locher cut the original route in half and nearly doubled the maximum gradient.

The technology for cog railways at the time would not support such an extreme ascent, so Locher came up with his own cog railway design that used two horizontal cogs on the outside of the centered track. This new design prevented the cogs from jumping off the track and also kept the carriages from toppling over in the high-wind environment. The original engine was a steam engine, and passenger rides began on June 4, 1889. It wasn't until 1937 that the service was electrified with overhead power supplies.

The railway is nearly three miles long and climbs over 5,000 feet through Pilatus massif to the Esel summit. The American Society of Mechanical Engineers named it a Historic Mechanical Engineering Landmark in 2001.

Above: *The Pilatus Railway climbs Pilatus, which is a mountain massif that is composed of several different peaks that are all nearly 7,000 feet tall.*

Below: *The views from the railway can even make the bravest souls hold on.*

A view of the Rhine River in the background.

The Drachenfels Railway is located in the North Rhine-Westphalia region of Germany and travels between the city of Konigswinter to the summit of Drachnefels Mountain. There are only three stations on this one-mile track, separated by a 900-foot difference in elevation. There are two terminal stations and an intermediate station that also functions as a passing loop for this single-track route. It uses the Riggenbach rack design for its cog system to overcome twenty-percent gradients up the mountain.

Opened in 1883 to transport tourists to the top of Drachenfels Mountain where the Drachenfels Castle remains, the Drachenfels Railway operates every thirty minutes during the day between March and October, while the months of November, January, and February provide less frequent departures. There is no service during the month of December.

The route operates a fleet of four four-wheel railcars with an overhead electric supply that was installed in 1953. Steam trains were used for many decades, and could still be seen in use after electrification until 1958 during high-traffic times of the year. In 1958, the derailment of a steam locomotive killed seventeen people and ended the use of steam engines on the line.

Above: *The train cars of the Drachenfels Railway were modernized in 1999 by the Swedish Locomotive and Machine Works Company.*

Left: *The Drachenfels Railway pictured with the Schloss Drachenburg, a nineteenth-century, palace-style private villa, in the background.*

The Semmering Railway at the Katle Rinne Viaduct

THE SEMMERING RAILWAY

The Semmering Railway of Austria crosses over the Semmering Pass of the eastern part of the Northern Limestone Alps, travelling between the Lower Austrian mountain town of Gloggnitz to Murzzuschlag in the state of Styria, Austria. It is considered to be the world's first mountain railway built on a standard gauge track. The track ascends nearly 1,500 feet at an average gradient of two percent for nearly sixty percent of the route's length. Along the service's route there are fourteen tunnels, sixteen viaducts, one-hundred stone arch bridges, and eleven small iron bridges—all evidence of the difficult terrain that had to be overcome to build this railway.

Albanian engineer Carl von Ghega designed the Semmering Railway, which was built between 1848 and 1854 by nearly 20,000 laborers. The philosophy of design applied by Ghega was quite progressive for its day. In an age when the brute force of dynamite was relied upon to clear the way for tracks, Ghega approached the Semmering Railway as a project in landscape gardening. He wanted the train to mesh with its surroundings and not impose on nature. This unique approach paved the way for the region to become more and more of a tourist destination. Even many of the stations along the line were built with the wasted stone that came from boring the tunnels along the route.

The Semmering Railway has been in continuous operation for over 160 years. It is one of the most beautiful railways passengers can book passage on. It was declared a UNESCO World Heritage Site in 1998.

***Above:** The Limestone Alps of Central Europe run parallel to the Austrian Central Alps but are a different range altogether.*

***Left:** Many of the viaducts along the Semmering Railway are double tiered for use by both cars and locomotives.*

THE JUNGFRAU RAILWAY

The Jungfrau Railway can transport up to 230 passengers per train. In 1997, the railway served more than 500,000 people for the first time in its history.

Switzerland's Jungfrau Railway is a railway unlike any other. Departing from the Kleine Scheidegg Train Station at 6,762 feet above sea level, the Jungfrau Railway ascends the Bernese Alps to its terminal station at Jungfraujoch, a prominent mountain pass between the near 14,000 foot peaks of Jungfrau and Mönch. Located within the high-altitude Top of Europe building at Jungfraujoch, the Jungfraujoch Station is the highest railway station in Europe at 11,332 feet above sea level. At the Top of Europe building, passengers can go to restaurants, peruse history exhibits, or explore the glacial ice caves of Ice Palace. From the Top of Europe building you can take an elevator to the summit of Jungfrau to the Sphinx Observatory, one of the world's highest observatories that houses laboratories, meteorological observation stations, and scenic terraces. Climbing 4,570 feet in a span of nearly five miles through the Jungfrau Tunnel, bored through Mönch and Eiger Mountains, the railway has five stops in which the passengers can disembark at underground stations like the Eismer Station, where passengers can view the Ischmeer Glacier from windows installed into the sides of the mountain.

The idea of building a railway to the top of Jungfrau first gained steam in the 1860s, but did not gain momentum due to financial barriers. In 1894, a concession to build a railway to the summit of Jungfrau was given to the industrialist Adolf Guyer-Zeller, and work began just two years later. The railway progressed quickly, but not without its dangers. Tracks made it to Eigergletscher Station, the railway's last open-air station, in 1898. Six workers were killed in a explosion in 1899, and Guyer-Zeller died that year after a four-month worker strike, but construction continued. By 1905, the railway had reached the destination of its penultimate station at Eismeer. Nearly thirty tons of dynamite exploded in 1909 at the Eigerwand Station, but that was the last major tragedy before success. The boring work concluded when the crews broke through the glacier in February of 1912, and the Jungfraujoch Station officially opened in August of that year. Although the original plan called for the railway to reach the Sphinx rock formation at the summit of Jungfrau with seven stations along the route, the achievement of creating a railway such as the Jungfrau is enough of an achievement.

After the railway was completed, the complex of research facilities, observatories, exhibits, and restaurants were built out from the underground station onto the Bernese mountainsides. The "house above the clouds," the Top of Europe building, was built in 1924, while the Sphinx Observatory was not finished until 1937. The Jungfrau complex also acquired a snowblower that year, which permitted year-round operation. The Jungfrau Railway is a rack railway built to be powered by electricity. The railway reaches gradients of twenty-five percent and travels through three tunnels for eighty percent of the trip's duration—the longest tunnel, the Jungfrau Tunnel, being nearly four and a half miles long. Along with the Aletsch Glacier, the entire Jungfrau-Aletsch area was declared a UNESCO World Heritage Site in 2001.

The Jungfrau Railway at the Kleine Scheidegg Station.

The train operates on a regenerative power system that generates electricity on the descent that is distributed back to the grid. The system produces nearly half of the power needed to ascend the mountain.

The Cinque Terre Line at the Manarola Station. Manarola is the second smallest town in Cinque Terre. The cornerstone of the San Lorenzo Church dates back to 1338.

THE CINQUE TERRE LINE

The Cinque Terre Line travels along the Mediterranean Coast through the five historic and isolated Italian Riviera towns of Cinque Terre. Completely free of automobiles and inaccessible by road, the towns that make up the Cinque Terre region, Riomaggiore, Manarola, Corniglia, Vernazza, and Monterosso, all have histories that date back to the Middle Ages. Travellers can get to this region by rail with a connection at the La Spezia Transit Station in the Laguria region of Northern Italy, which is just minutes by rail to this colorful enclave filled with tradition.

Conceived under a plan to connect the Ligurian city of Ventimiglia with the Tuscan city of Massa, the Cinque Terre Line was given concession in 1860 by royal law. Using boats to haul in supplies for the twenty-three bridges and fifty-one tunnels needed in areas inaccessible by land, the twenty-seven miles of track was completed in the mid-1870s.

Traveling northwest from La Spezia, the Cinque Terre Line enters the Cinque Terre National Park along the Mediterranean Coast and makes its first stop at Riomaggiore. With just minutes between stops, the train then works its way through tunnels to the towns of Manarola, Corniglia, and Vernazza before it reaches the terminal station of Monterosso. If you're only using the train to gain access to this secluded region, you can also travel between the towns on foot by way of the Azure Trail. The Cinque Terre region is a UNESCO World Heritage Site.

The Cinque Terre region has been inhabited for centuries, with houses and buildings being built into the hillsides with the use of an extensive terrace system.

Left: *The scenery of this region is breathtaking, but such landscapes can be hard to live in. The area is prone to landslides and flooding when storms come through the region.*

Right: *The station at Vernazza. The first records of Vernazza date back to 1080, when the town was a fortified naval base for the noble Obertenghi family to defend the coast from African pirates.*

Nearly ninety-five percent of the Harz National Park is covered with forest.

***Above:** The Brocken Railway climbs 1,965 feet to the summit of Brocken. The mountain is often associated with witches, devils, and the occult in German folklore, and is known for a strange optical illusion known as the Brocken Specter.*

One of the three tourist trains that make up the Harz Narrow Gauge Railways network in the Harz Mountain range of northern Germany, the Brocken Railway leaves from the Drei Annen Hohne Station in Wernigerode, Germany, and travels along the Bode River to the summit of Brocken, the highest peak in the Harz Mountain range. The Harz Narrow Gauge Railway Company was formed after World War II in the merger of two previous railway companies. The company also operates the Harz Railway and the Selke Valley Railway in the Harz region. Operating a large number of steam locomotives on nearly eighty-seven miles of steeply graded track in the region, the nearly twelve-mile Brocken Railway is by far their most popular route.

Plans for the railway first surfaced in 1869, but it was not until 1895 that the plan was accepted, the construction permit issued, and the land was allocated. Work between the Drei Annen Hohne and Schierke Stations was completed in 1898, and construction to the summit of Brocken from Schierke continued after that. During World War II, the route was suspended due to damaged tracks from bombing in the fortified Harz region. Service opened again in 1949, but not for long. After the completion of the Berlin Wall in 1961, the Brocken was considered to be a part of the "out-of-bounds" area and was not accessible to the public. The train continued to operate in those days, but only as a freight service to ship coal, timber, and oil up the mountain to East German Border Troops and Soviet soldiers who were stationed at the Brocken military base. After Germany's reunification, the government questioned the viability and value of the Brocken Railway and considered closing it, but with the support of rail enthusiasts and politicians, the railway was saved from abandonment. Plus, the train was needed to haul away the obsolete military equipment from the former military facilities at the summit.

The train was renovated and reopened to the public in 1991 with two historic steam engines operating as the main locomotives. Operations run year round, with six pairs of trains running daily during the winter and increased service in the summer. It takes about fifty minutes to reach the summit on the fastest train, operating at an average speed of twenty-four miles per hour. The maximum gradient the route faces is close to three and a half percent.

***Right:** The Brocken Railway only uses steam locomotives.*

THE PIKES PEAK COG RAILWAY

Diesel railcars built by the Swiss Locomotive and Machine Works company in 1975 ascending to the summit of Pikes Peak.

Pikes Peak in a registered National Historic Landmark and is named after Zebulon Pike, an American explorer who was unable to reach the summit of the mountain in 1806 after two-days of foodless hiking through waist-deep snow.

Because of the peak's high elevation and polar climate, snow can be expected at any time of year at Pikes Peak.

Unlike the many train routes in Colorado that were built for mining and other causes that promoted Westward Expansion, the Pikes Peak Cog Railway was built solely for the entertainment of tourists. Although it has had a long history, the fate of this railway is unknown. It is the highest railway in North America, starting from the Manitou Springs Train Station at 6,412 feet above sea level to the summit of Pikes Peak at 14,115 feet above sea level. The route is almost nine miles long with an average gradient of sixteen percent, which is surmounted by an Abt rack system.

Zalmon G. Simmons, of Simmons Beautyrest Mattress Company fame, founded the railway in 1889. At that time, limited service was provided to the Halfway House Hotel, located in Ruxton Park, Colorado, about halfway between Manitou Springs and Pikes Peak. By 1891, the railway reached the summit, and service was opened to the public. The railway used steam locomotives for the first few decades of service, which can be incredibly arduous and difficult work for a railway that operates on such a gradient. Gasoline and diesel powered railcars were purchased for the railway by 1938, but steam-powered service was still provided until the 1960s. By the 1970s, the railway was required to increase its capacity because of the number of tourists that were attracted to the destination, which led to the railway acquiring four 214-passenger railcars in 1976. In addition to those railcars, the Pikes Peak Cog Railway also holds four smaller Swiss-built railcars, four GE locomotives, a snowplow, one steam locomotive, a Winter-Weiss "streamliner" coach, and an original Wasson wooden coach.

The Pikes Peak Cog Railway tended to provide service six to eight times a day during the peak months of the summer, usually closing between mid-December and mid-March due to the amount of snow on the mountain. In 2006, the railway started operating year round when the snowplows were able to clear the tracks. But the service was put on hold in the winter of 2017, when scheduled maintenance uncovered larger maintenance issues that were too expensive to complete. It is said that the entire infrastructure of the railway needs to be replaced, including the passenger cars, rails, and rail ties. The Manitou Springs City Council gave tax incentives to the railway in 2018 to help with the repair costs. At this point, it is estimated that the railway will reopen in 2020.

The Centovalli Railway crossing the Melezza River.

THE CENTOVALLI RAILWAY

The Centovalli Railway is a transnational route that runs between Domodossola, Italy, and Locarno, Switzerland, on thirty-two miles of track. Traversing the mountainous and valley-ridden terrain, the rail is strategically placed to easily travel across the Alps between Italy and Switzerland. There are twenty-two stations along the route with the Italian/Swiss border lying between the stations of Ribellasca, Italy, and Camedo, Switzerland. The railway crosses deep ravines over high viaducts, passes through numerous valleys lined with thick forests, and stops at many small mountain villages in the Swiss and Italian countryside.

Construction of the railway began in 1918 after Italian and Swiss diplomats signed a convention to create a connection between their two countries. Tracks from a former railway between Locarno and Bignasco were used on the eastern, or Swedish, end of the Centovalli Railway, which also set the standard gauge that would be used for the rest of the railway. Many thought that the railway would be under-utilized after automobiles became ubiquitous in the mid-twentieth century, but the railway has continued to play a big role in transporting tourists and residents between Domodossola, Italy, and Locarno, Switzerland.

The Centovalli Railway is operated by both Swedish and Italian rail companies and provides first, second, and panoramic coach tiers of service. As you pass through the land of a "Hundred Valleys," you will pass by some of the most beautiful views Europe has to offer along with many villages that time seems to have forgotten.

Top right: *A steel bridge outside of the Swiss town of Intragna.*

Bottom left: *The train at the Camedo Station, which is the closest Swiss station to the Italian border.*

Locomotive #2 was originally built in 1930 for the Eastern Province Cement Company in Port Elizabeth, South Africa. It was bought by the Brecon Mountain Railway in 1990, fully restored, and put into operation in 1997.

Founded in 1977 as a tourist railway traveling between Pant, Wales, to the Brecon Beacons National Park along the Pontsticill Reservoir, the Brecon Mountain Railway is a privately owned operation that is run by Tony Hills, who also operates and owns the Llanberis Lake Railway (see pages 38–39). Hills bought a five-mile stretch of track along the old Brecon & Merthyr Railway after he established a base for his locomotive collection at the Llanberis Lake Railway in 1970. Construction began along the abandoned railway's track bed toward the new terminal station of Torpantau in 1978. Along with the track bed, several bridges were repaired and the station at Pontiscill was renovated. Passenger service began in 1980 with one steam engine in service.

The Brecon Mountain Railway follows the old track bed of the Brecon & Merthyr Railway, which was built in 1863 to connect with the Dowlais Railway. The Brecon & Merthyr Railway was to provide the Dowlais area access to the shipping canal near Brecon to transport iron ore. The area is a very isolated area, and methods to ship goods out of the Dolwais are very valuable. The Monmouthshire & Brecon Canal was opened in 1800, allowing goods north of the Brecon Beacons to ship out of the area easily, while everything south of the Brecon Beacons had a more difficult time crossing the rugged terrain. The Brecon & Merthyr Railway provided an easy way to transport those goods north. By the 1920s, iron production in the Dowlais area had declined considerably, and for the next four decades passenger and freight service would incrementally be withdrawn from the route, section by section. The section of track between Merthyr and Pontsticill, the general section of track that the Brecon Mountain Railway opened on, was closed under the Brecon & Merthyr Railway's management in 1961.

Today there are four stops along the Brecon Mountain Railway. The Pant and Torpantau Stations serve as the south and north terminuses of the line, while the station at Pontsticill and Dolygaer Stations are located on the south and north ends of the Ponsticill Reservoir. The Dolygaer Station is not an actual stop on the line, but works as a passing loop for the narrow-gauge route. The railway's northern terminal is just south of the Trapantau Tunnel, the highest railway tunnel in the UK that was once known as the Devil's Tunnel due to its 666-yard length.

Above: *The Brecon & Merthyr Railway had the unfortunate nickname of "The Breakneck & Murder Railway" due to the alarming number of accidents that occurred on the railway. These accidents were often due to the steep gradients that the railway traversed.*

Left: *The railway has twelve different locomotives in its rolling stock, although not all of them are operational.*

THE RIO GRANDE SCENIC RAILWAY

The Rio Grande Scenic Railway operates both steam and diesel engines.

Above: *The Rio Grande Scenic Railway just outside of La Veta with the Spanish Peaks behind it.*

Located 200 miles south of Denver, Colorado, in the larger area of the San Luis Valley, the Rio Grande Scenic Railway opened in 2006 as a tourist train. Travelling east between Alamosa, Colorado, and Walsenburg, Colorado, located in the smaller La Cuchara Valley below the Spanish Peaks, the Rio Grande Scenic Railway travels for sixty miles along what was originally the San Juan Extension of the Denver & Rio Grande Western Railroad.

The Denver & Rio Grande Western Railroad was incorporated in 1870 with the intent of working its way south along the Rio Grande toward El Paso, Texas. The narrow gauge railway reached Colorado Springs by 1871, moved south toward Pueblo, and made it to the Veta Pass and into Alamosa by 1878. The Denver & Rio Grande Western Railroad is credited with opening the San Luis Valley up to the rest of the world, allowing for the area's rich resources to be easily transported out of the area and enter into the world's economy. The city of Alamosa was literally built in one day as buildings were transported into town on the railway the day the station opened. Alamosa was the narrow-gauge hub of North America, with more narrow-gauge tracks operating out of the city than any other city. Though, the original narrow-gauge tracks that operated between La Veta and Alamosa were replaced with standard gauge in 1899 to compete with larger freight companies.

The Rio Grande Scenic Railway is owned by the San Luis & Rio Grande Railway, which bought 154 miles of Denver & Rio Grande Western Railroad tracks in 2003. The San Luis & Rio Grande Railway operates three lines radiating out of Alamosa. The Rio Grande Scenic Railway runs along tracks from Alamosa to a junction with the Union Pacific in Walsenburg, Colorado. Another train operated by the San Luis and Rio Grande Railroad runs south from Alamosa to Antonito where it meets with the Cumbres & Toltec Scenic Railroad (see pages 76-77).

Below: Icy tracks crossing the Veta Pass (9,220 feet) in the Sangre de Cristo Mountains. The Denver & Rio Grande Western Railway's narrow-gauge track used a different route to cross over the pass, but when the track was converted to standard gauge, the route was moved about seven miles southeast of the old route.

The section of track known as the Devil's Nose was repaired in 2001.

THE NARIZ DEL DIABLO RAILWAY

The Nariz del Diablo Railway, or the Devil's Nose Railway, is a short and scenic tourist train that runs along tracks of the Ecuadorian Railway Company between the cities of Sibambe and Alausi, Ecuador. The journey between the cities is relatively short, but the elevation gain is tremendous. In the seven miles between stations, the train traverses near-vertical mountainsides to ascend nearly 1,600 feet using a series of switchbacks. The train negotiates the switchbacks by moving past the switchback junction and then rolling in reverse to pass the next section's junction to then move forward toward the next section's junction. The Ecuadorian Railway Company provides service across the country, connecting the harbor of Guayaquil, via the city of Duran, to the capital city of Quito in the Andean foothills. The railway company manages the largest amount of infrastructure in the country, which is no easy task considering the difficult terrain presented by the Andes Mountains.

Built between 1897 and 1908, the construction of the Ecuadorian Railway Company took a tremendous toll on those who worked on it, especially the section of track that came to be known as the Devil's Nose. Building on the coastal plains or on the central plateau was easy, but managing to build tracks that created a transition between those two areas was quite difficult. Surveyors finally found a suitable area to make the ascent at Devil's Nose, then called the "Condor's Aerie." The name was changed due to the amount of deaths that occurred during the railway's construction. Nearly 3,000 Jamaicans and 1,000 Puerto Ricans were brought into the country as laborers, and it is estimated that nearly 2,000 laborers died during the construction of the Devil's Nose section.

By the late 1990s, nearly ninety percent of the rail was in disrepair and had been abandoned. Landslides, mudslides, flooding, and a variety of damage have plagued the tracks. By 1998, almost the whole railway was abandoned, except for small sections that operated tourist trains. The Devil's Nose Train survived these hard times, and by 2008, the president of Ecuador began a restoration project that would completely revamp the entire railway. By 2013, the whole route between Guayaquil to Quito was reopened to passenger traffic.

The train travelling across the sharp switchbacks that ascend the Devil's Nose.

It is said that passengers were once allowed to ride on the roof of the train to get a better view of the scenic Devil's Nose.

The route crosses enormous grasslands that occupy the Tibetan Plateau north of the Himalayan Mountains.

The Qinghai-Tibet Railway probably holds a world record for holding the most world records for a railway. Being the highest railway in the world, with the highest railway station and tunnel in the world, while also being the first railway to connect the autonomous region of Tibet to any other province in China, makes the Qinghai-Tibet Railway a remarkable route. Traversing nearly 2,000 miles between Xining, in the province of Qinghai, China, and Lhasa of the Tibet Autonomous Region, the railway has 340 miles of track laid on permafrost and contains 675 bridges. Nearly eighty percent of the line between Golmud and Lhasa operates at least 13,000 feet above sea level, requiring emergency oxygen supplies for all of its passengers and doctors on board at all times to treat altitude sickness.

The section of track between Xining and Golmud was finished in 1984, but the section between Golmud and Lhasa would not be completed until 2006 because of the difficulties presented by building on permafrost. Construction on the Golmud-Lhasa section began in 2004 from both directions. The Tanggula Pass Station, the highest train station in the world at 16,627 feet above sea level, was completed in 2005. The railway was a part of the China Western Development program that looked to develop the western section of the country, which has historically been less industrialized than the eastern part of the country. The railway has helped reduce the cost of imported goods into Tibet. Because of the limited industrial capacity of Tibet and the limited transportation into the area, the railway is able to bring more goods into the country faster and cheaper than other modes of transport.

This epic trip across the Tibetan Plateau is unlike any other train route in the world. Travelling across extremely high elevations through super isolated areas, the Qinghai-Tibet Railway gives access to some of China's most scenic natural areas, including the Tanggula Pass and Mountains, the World Heritage Site of the Kekexili Grasslands, the Nagqu Grasslands, Yuzhu Peak, and Qinghai Lake.

Top right: *One of the many land bridges used along the route to cross the permafrost regions in high-altitude Tibet.*

Left: *The Qinghai-Tibet Railway in front of the Tanggula Mountains.*

THE SETTLE-CARLISLE LINE

The Settle-Carlisle Line at the Ribblehead Viaduct in North Yorkshire.

A part of the United Kingdom's National Rail network, the Settle-Carlisle Line runs between the Yorshire Dales at Settle and the Carlisle Station near the Scottish/English border in Cumbria. The Midland Railway opened the line in 1875 to transport goods north and passenger service opened in 1876. It cost nearly £3.6 million to build the railway, which was nearly fifty percent more than the expected cost of the rail. From the very beginning, the railway struggled to compete with rival railways that provided service from England into Scotland. The London-Glasgow Line of the powerful London & North Western Railway could not keep pace with the number of trains that the Midland Railway's Settle-Carlisle Line provided at first, but after its merger with the London & North Western Railway, the Settle-Carlisle Line seemed to be a bit redundant, especially after the nationalization of the country's railways in 1948.

A crew of nearly 6,000 land navigators, or navvies, who endured the harsh weather and isolation of the region, built the Settle-Carlisle Line. Camps were formed in these isolated regions by the workers, and they began to grow large enough to provide postal services and schools. Many workers died during construction, although the official numbers are still unknown. Nearly eighty people died at the Batty Green camp, near the Ribblehead Viaduct today, from a smallpox outbreak.

The line was engineered for express service, so local stops were not accounted for at first. But as time went on many local stops were added to the route. There are nearly fourteen tunnels and twenty-two viaducts

along the route, including the twenty-four arch Ribblehead Viaduct, which was necessary in order to build tracks along the area's bog-like terrain.

By the 1980s, the Settle-Carlisle Line was nearly obsolete and slated for closure. The price needed to restore the viaducts, tunnels, and track was much too high considering the low numbers in ridership. Many stations along the line had been closed since the 1950s, and the remaining stations in operation were notified of the line's closure in 1984. Outrage from locals came quickly. Many claimed that the line was not an insignificant branch on the rail network, but a main line that many people relied on. The committee of supporters even uncovered that National Rail was conspiring against the line, claiming that the cost of repairs was much lower than the rail network said it would be. They even found evidence that the rail network was diverting traffic from the line to affect the line's passenger numbers. The protest worked, and ridership went from 93,000 a year in 1983 to 450,000 a year by 1989. The government decided to keep the line open and began the repair work that was needed. Since then, the ridership and infrastructure has increased dramatically. Unfortunately, the line has been a victim of temporary closures in 2015 and 2016 due to massive amounts of flooding and landslips along the route.

Top: *The Ribblehead Viaduct, also kown as the Batty Moss Viaduct, crosses the River Ribble, the only river in Yorkshire that flows westward.*

Bottom left: *The Settle-Carlisle Line at the Garsdale Viaduct on the Cumbria and North Yorkshire border.*

Bottom right: *The Settle-Carlisle Line passing one of the former construction camps that were built by those who constructed the line.*

Snowdonia National Park is 823 square miles in area and was the first of three national parks dedicated in Wales in 1951.

THE SNOWDON MOUNTAIN RAILWAY

The Snowdon Mountain Railway is a rack and pinion railway that travels to the summit of Snowdon Mountain, the highest peak in Wales. With a track length of nearly five miles starting from Llanberis and reaching its terminus at the summit, the Snowdon Mountain Railway has operated for more than a hundred years as a tourist attraction, carrying nearly 130,000 passengers annually. The railway travels throughsome of the United Kingdom's harshest weather conditions, with regular high velocity winds, drastic temperatures, and extremely wet conditions. The Snowdonia National Park is known to be one of the wettest regions of the United Kingdom, with some parts of the park receiving an average of 176 inches of rain a year. Harsh weather is even known to cancel scheduled service on the route, and keeps the railway closed between the months of November and March.

The idea for the Snowdon Mountain Railway first emerged in 1871, but a local landowner, Mr. Assheton-Smith, who thought that the railway would ruin the area's natural scenery and isolation, opposed it. After two decades, Assheton-Smith gave way to the idea and conceded his land to the railway company. No compulsory purchase order had to be issued by the government, and no Act of Parliament was needed for construction approval since it was all built on private land. The railway was constructed between December of 1894 and February of 1896. Laying the trackbed was about half done by early 1895, and the track installation began in the summer of 1895. The first train to reach the summit was in January of 1896. Days before the opening, on April 4, 1896, a contractor was descending the mountain in the train, hit a boulder, and derailed the train. The workers mounted the railcar back on the tracks and continued on their way. Two days later, on the opening day of the route, two trains were dispatched to the summit, but on their way down, the first train derailed due to the car's weight, and the second car, not hearing of the first's derailment, continued down, derailed too, and ran into the first car. Only one person died during the incident. The train was closed after the incident and reopened in September of 1896.

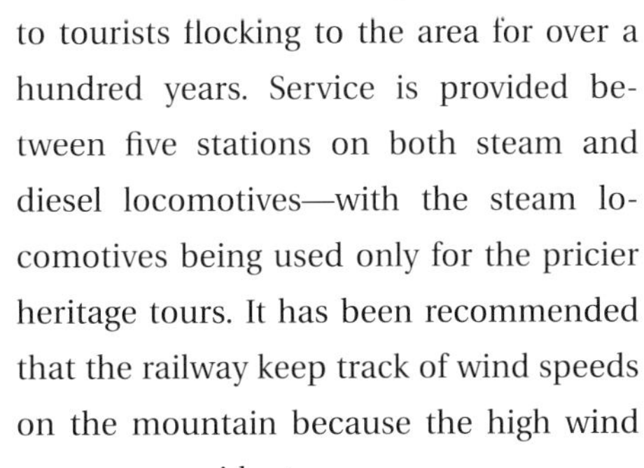

Since then, the train has provided service to tourists flocking to the area for over a hundred years. Service is provided between five stations on both steam and diesel locomotives—with the steam locomotives being used only for the pricier heritage tours. It has been recommended that the railway keep track of wind speeds on the mountain because the high wind can cause accidents.

Top right *The summit of Snowdon Mountain is 3,560 feet above sea level.*

Bottom right: *Sheep, polecats, otters, goats, and pine martens are some of the types of wildlife you can see in Snowdonia National Park.*

THE FFESTINIOG RAILWAY

The Ffestiniog Railway is a part of the Great Little Trains of Wales initiative, which seems quite accurate considering the diminutive size of this train compared to behemoth diesel locomotives we commonly see today.

Yet another narrow-gauge heritage train route found in the Snowdonia National Park in Wales, United Kingdom, the Ffestiniog Railway has more history than any of the other heritage rails that also occupy the area. Owned by the Ffestiniog Railway Company, the oldest operating railway company in the world, which also operates the Welsh Highland Railway, the Ffestiniog Railway travels for thirteen miles between Porthmadog and the slate-mining town of Blaenau Ffestiniog. The Ffestiniog Railway operates along four stations through mountainous areas and forests, across the dike of Traeth Mawr, the Cob, and past numerous slate quarries.

The railway was constructed between 1833 and 1836, but locomotives were not used for the first few years. Instead the railway had a gradient that descended from the slate quarries at Blaenau Ffestiniog, so loaded railcars could be transported to the port by gravity alone. The empty railcars were then transported back up the hill by horse. Steam locomotives were bought, modified for the narrow-gauge tracks, and put into operation by 1863. These locomotives allowed for longer slate trains in order to keep up with the quarry's growing production. Steam trains also introduced the passenger service to the line. But passenger service and the use of gravity-operated trains ceased in 1939, after a decade or so of tapering slate production. Slate traffic ceased in 1946, and the railway was nearly abandoned when local rail enthusiasts began talking about the route's restoration, which began in 1954. The railway was in bad shape, and new trackbeds were built to create a deviation away from the Tanygrisiau Reservoir, which now flooded a section of tracks after the construction of the Ffestiniog Pumped Storage Scheme in 1954. Construction took much longer than expected, and the railway was not officially reopened until 1982.

Since then, the railway has carried a yearly average of 200,000 passengers. It is the second most popular tourist destination in Wales, following the medieval fortress of Caernarfon Castle. Since its reopening, the railway has received many subsidies and grants from various institutions because of its historical importance to the area. Today it is promoted as a part of the Great Little Trains of Wales initiative, which promotes the upkeep of the numerous narrow-gauge railways that are found in Wales.

Top right: *A picture of the railway with both passenger and slate services running in 1900.*

Bottom left: *The Cwmorthin Quarry was opened in 1810, connected to the Ffestiniog Railway in 1860, and completely abandoned by 1997.*

Trains operate between April and October, with some service during winter weekends and holidays.

THE NORTH YORKSHIRE MOORS RAILWAY

Built in stages in 1836, the North Yorkshire Moors Railway was once called the Whitby & Pickering Railway, which ran on the east coast of England in an attempt to increase traffic into and out of the dwindling port city of Whitby. Planners for the railway believed that opening a railway that connected the interior of the country to the port of Whitby would help revitalize the city's economy, which was crumbling around its traditional industries of whaling and shipbuilding. The railway's route was originally going to be used for the construction of a canal between the Pickering and Whitby, but after the success of the nearby Stockton & Darlington Railway, the plans were modified to construct a railway instead.

The railway opened both freight and passenger services in 1836, operated by horse-drawn carriages on a single track. It was the first railway to open in North Yorkshire. By 1845, the railway was absorbed into the expanding empire of the York & North Midland Railway, who soon began to convert the railway from horse-drawn carriages to steam locomotives. Whitby was soon connected to many other towns in the greater region, including Hull, Manchester, Liverpool, West Riding, and London, creating larger numbers of day-trip and holiday passengers on the line. By 1849, the Royal Mail service was running daily trains from York to Whitby, often the first train in the morning. Throughout the following decades, the Whitby & Pickering Railway was absorbed multiple times before the nationalization of the British railway system in 1948.

By the 1960s, the Whitby & Pickering Railway, then operated by British Railways, came under the scrutiny of the Beecher Axe, a report analyzing the utility and organization of the British Railway system that suggested the closure of fifty-five percent of Britain's train stations and thirty percent of its total track mileage. The track between Rillington Junction and Whitby was closed due to the report's recommendation, and it wasn't until 2007 that the North Yorkshire Moors Railway was granted permission to begin its operation on the track once again.

The North Yorkshire Moors Railway is the busiest steam heritage route in the world, carrying 355,000 passengers a year.

Although trains currently run between Pickering and Whitby, there is talk of reopening the eight-mile track that leads to Rillington Junction.

THE TRANS-HARZ RAILWAY

The Trans-Harz Railway stopped at its junction with the Brocken Railway.

Operated by the same company that operates the Brocken Railway (see pages 90–91), the Harz Narrow Gauge Railway Company, the Trans-Harz Railway is a thirty-eight mile track that crosses the Harz Mountains between Wernigerode and Nordhausen, Germany. It operates at an average speed of twenty-five miles per hour with a maximum gradient of three percent. Running north from Nordhausen, the Trans-Harz Railway parallels a standard gauge route, the South Harz Line, which runs between Nordhausen and Ellrich. Around the town of Ilfeld, at about mile six of the route, the train begins to work its way into the Harz Mountains. This section of track was the first section to be completed in 1897.

After the connection between Nordhausen and Ilfeld was completed, it was then extended to Netzkater and Benneckenstein in 1898. That same year, the section of track being constructed from the northern terminus of Wernigerode to Drei-Annen-Hohne was completed, while the connection between Drei-Annen-Hohne and Benneckenstein completed the whole route in 1899, the same year the Brocken Railway was completed. The Trans-Harz Railway has junctions with both the Brocken Railway at Drei-Annen-Hohne and the Selken Valley Railway at the Eisfelder Talmühle Train Station.

Top left: *The track gauge of the Trans-Harz Railway is 1,000 mm wide.*

Top right: *The Trans-Harz Railway departing from the Wernigerode Station.*

Bottom left: *The Trans-Harz Railway has two steam locomotives for operation, as well as diesel locomotives that help transfer goods along the line. The only freight service still operating on the line runs between the stations of Nordhausen and Eisfelder.*

The Darjeeling Himalayan Railway was declared a UNESCO World Heritage Site in 1999.

THE DARJEELING HIMALAYAN RAILWAY

The Darjeeling Himalayan Railway travels between the cities of New Jalpaiguri and Darjeeling in the state of West Bengal, India. Built in between 1879 and 1881, the railway is known for its six zigzag switchbacks and five loops along the fifty-five miles of its route. It is a UNESCO World Heritage Site. Scheduled passenger services run with diesel locomotives while tourist services run with heritage steam locomotives. The train ascends over 6,000 feet above sea level on an adhesion railway, with the help of various types of infrastructure to surmount steeper areas.

After the success of the 1878 railway between Siliguri (not far from New Jalpaiguri) and Calcutta, the Eastern Bengal Railway proposed to build a railway between Siliguri and Darjeeling to replace the *tonga* services transporting goods and people on Hill Cart Road. The proposal was accepted and construction began quickly. The route follows the old Hill Cart Road for a good portion of its route. Many of the cart road's sections were too steep for a locomotive to traverse, so four loops and four zigzag switchbacks were installed to ease the steep gradient between Sukna and Kurseong. By 1910, it is reported that the railway was carrying 174,000 passengers a year, along with 47,000 tons of goods.

By the mid-twentieth century, the railway was facing competition from bus services that now ran along Hill Cart Road, so the railway constructed the Batasia Loop to create easier ascents into Darjeeling.

The railway follows Hill Cart Road for much of the way, which can cause some problems. These days, the road is surrounded by neighborhoods and houses, making the train look more like a metropolitan service than an intercity route. Because of the densely crowded areas it passes through, there are major risks to all of the pedestrians that find themselves walking near or on the tracks. The trains are equipped with extremely loud horns that are blaring almost constantly in these areas. A prime example of this can be seen at the market in Kurseong.

Top right: *The Darjeeling Himalayan Railway passing through the Kurseong Market.*

Bottom Left: *The train next to the old Hill Cart Road. The railway has a preserved gauge of two feet.*

THE MATHERAN HILL RAILWAY

The Class NDM-1A diesel locomotive that was built in 2006 is one of six diesel locomotives in operation on the Matheran Hill Railway.

Still on the list to be elected as a UNESCO World Heritage Site, the Matheran Hill Railway is a narrow-gauge heritage railway that connects the town of Neral to the Western Ghats railway station of Matheran. The route is thirteen miles long and travels through the dense forests of the Western Ghats, a mountain range that is itself a UNESCO World Heritage Site and known as one of the eight best regions for biological diversity in the world. Like the Darjeeling Himalayan Railway, it has a two-foot wide preserved gauge.

The railway was built as a private investment between 1901 and 1907 by Abdul Hussein Adamjee Peerbhoy for $160,000. Adamjee Peerbhoy frequently visited Matheran, which lead him to the idea of constructing a railway to the top of the mountain to make the trip easier. The route was surveyed in 1901 and construction began in 1904, with the route being completely finished by 1907. The train reaches a maximum gradient of five percent and can only travel at seven and a half miles per hour.

From Neral, which is not far from Mumbai, the track runs along a standard-gauge track until it reaches Hardal Hill where it begins to ascend Matheran Hill. The track seems to level off a bit before it begins to climb Mount Barry, where a horseshoe-shaped embankment was built to help the train make it up the steep ascent. At the top of Mount Barry, it goes through One Kiss Tunnel before it reaches two zigzag switchbacks that help the train negotiate through the deep sections before Panorama Point. It takes a long two hours and twenty minutes to make it up the hill, but the operator of the railway plans on reducing that time to one hour and thirty minutes in the next few years.

Top right: *The area is prone to flooding and landslides due to the steep mountains and the amount of precipitation the area receives every year.*

Bottom left: *The Matheran Train Station covered in fog. On the left, you can see the O & K manufactured steam engine that was made in 1905 for the railway.*

The Konkan Railway connects three states in India: Maharashta, Goa, and Karnataka.

THE KONKAN RAILWAY

Before the completion of the Konkan Railway, the major economic centers and port cities of Mangaluru and Mumbai, India, were not directly connected by a rail network, leaving passengers and goods to travel inland on various lines to make a connection between the two cities. The first section of track, built between Mangaluru and Udupi, was opened in 1993, and the whole route was not completed until 1998. In its entirety, it is 459 miles long and it usually operates at about seventy-five miles per hour—although it was built to withstand trains travelling at ninety-nine miles per hour. It is a single-line track that operates with diesel locomotives.

The construction of the railway was filled with much controversy. Although planning had begun for the railway in 1988, surveying was not yet complete. Nearly half of the route, lying in the state of Maharashtra, was unsurveyed, and the planners had no idea what type of terrain they were working with. The Ghats of western Maharashtra proved to be exceptionally difficult. The route needed a total of 2,000 bridges and ninety-one tunnels to be built in order to traverse the difficult terrain. Flashfloods, landslides, and tunnel collapses all affected the construction progress as well. Controversies around the route also began to upspring in the state of Goa, where many activists said that the proposed route had various ecological, cultural, and economic implications for the surrounding area and population. They believed the railway would produce an unfair burden on their state. Despite all the controversies, the railway progressed quickly.

Thc Konkan Railway operates both passenger and freight services between the two major port cities. There are a variety of express trains that passengers can take between the route's sixty-seven stations, while freight services have not been as popular since the route's inauguration. Although it has only been open for twenty years, the railway has many problems resulting from the extreme weather and geography of the region. The monsoon season of 1998 washed away portions of track, flooded other portions, and disrupted services for some time. A landslide in 2003 caused an express train to derail at the entrance of a tunnel, leading to the deaths of fifty-one passengers. Just a year later, another express train was derailed after colliding with a boulder on the track. The train fell off a bridge and left twenty people dead. This nearly new route has done wonders for the economy of India, but it has a few kinks in its infrastructure that still need to be addressed.

The Konkan Railway is a single track railway, but there is talk of doubling the track where the trackbed is wide enough.

Many precautions are being implemented along the route to prevent landslides and boulders from covering the tracks.

THE NILGIRI MOUNTAIN RAILWAY

Built by the British in 1908, after forty-five years of bureaucratic indecision, the Nilgiri Mountain Railway is a twenty-eight mile long service between the cities of Mettupalayam and Ooty of India's southern state of Tamil Nadu. It is India's only rack railway. Although it is operated almost exclusively with steam locomotives, diesel locomotives are used between the stations Coonoor and Ooty. It takes nearly five hours for the train to make the ascent to Ooty, while another four hours are needed to make it back to Mettupalayam. The train traverses over a hundred curves, sixteen tunnels, and 250 bridges at a maximum speed of nineteen miles per hour on flatter ground and nine miles per hour while it's climbing.

Originally opened as a passenger service between Mettupalayam and Coonoor in 1899, the railway was extended to Fernhill in September of 1908 and then to its new terminus of Ooty in October of 1908. It was operated by the Madras Railway, a railway that was essential in the development of railways in southern India, until it was purchased by the Southern Indian Railway. In 2005, it was designated a UNESCO World Heritage Site, joining the Darjeeling Himalayan Railway (see pages 114–115) as one of the "Mountain Railways of India."

The train ascends more than 6,000 feet among its thirteen stations at a gradient averaging around four percent. The maximum gradient the train reaches is eight percent outside of the station at Kallar. Many stations along its route, like the Adderly and Runnymede Stations, are used only as water stations to resupply the train's steam engine.

Top: *The Nilgiri Mountain Railway operates on standard-gauge tracks.*

Center: *The Nilgiri Mountain Railway has been featured in numerous Bollywood films, including the Indian romantic film* Dil Se.

Bottom: *A train outside of the Coonoor Train Station, one of the original terminal stations when the first phase of construction was completed in 1899.*

THE MONTSERRAT RACK RAILWAY

The rack railway ascending up Montserrat from the station at Monistrol Vila.

The Montserrat Rack Railway travels for just over three miles between the city of Monistrol de Montserrat in Catalonia, Spain, to the eleventh-century Benedictine mountaintop monastery of Santa Maria de Montserrat at the top of the Montserrat Range. The service uses both conventional adhesion and rack systems to ascend 1,804 feet at a maximum gradient of fifteen percent. It can operate at a total of eighteen miles per hour on the rack system and twenty-six miles per hour on the conventional adhesion track. From the station in Monistrol de Monsterrat the line travels along the adhesion track to the station of Monistrol Vila where a rack railway continues to the monastery. From the summit, two other rack railways can be taken: the Funicular de Sant Joan services passengers to the summit of Montserrat, while the Funicular de Santa Cova services passengers to a hiking trail that goes to the Santa Cova Cave.

Completed in 1892, the Montserrat Rack Railway successfully operated for many decades before the Aeri de Montserrat aerial cable car was opened in 1930. The Aeri de Montserrat took much of the Montserrat Rack Railway's customers, and the rack railway fell on hard times. An accident in 1953 and poor financial earnings throughout the 1950s led the railway to its closure in 1957. But as the Santa Maria de Montserrat grew in popularity as a tourist destination, the aerial cable car was unable to keep up with the demand. After decades of being closed, the Montserrat Rack Railway was restored for use in 2003.

The train approaching the monastery.

The Benedictine monastery at the top of Montserrat is believed to be the home of the Holy Grail according to some Arthurian Legends. Along the mountain range, you can find numerous caves that were once used by reclusive monks.

Departing from the Monistrol de Montserrat Train Station, the Montserrat Rack Railway passes through the 524-foot La Foradada Tunnel, which is directly followed by crossing the River Llobregat on the Pont del Centenari Bridge. The train then travels upon the banks of of River Toruguer until it reaches the Monistrol Vila Station. From there, passengers transfer to the rack railway system that passes through the Tunnel of Ángel, built at the time the original railway was built. Two more bridges and the Tunnel of Apóstols complete the line as it approaches Montserrat. The Funicular de Sant Joan and Funicular de Santa Cova depart from the Montserrat Station. The Funicular de Santa Cova descends down the range a bit to a hiking trail that goes to the Santa Cova Cave, a cave described in Spanish folklore as a place where the image of the Virgin Mary was seen by shepherds in 880, and the Funicular de Sant Joan proceeds to the summit of Montserrat where passengers can take in breathtaking views of the range.

The Taieri Gorge Limited passing over one of the many viaducts in the Taieri Gorge.

THE TAIERI GORGE LIMITED

Formerly known as the Taieri Gorge Railway, the Taieri Gorge Limited is an excursion train that departs from Dunedin, New Zealand, through the Taieri River Gorge to Pukerangi, a small town in the Otaga region of New Zealand's South Island. This scenic route, that travels across the Taieri Plains and then down a deep and narrow ravine carved by the Taieri River, once called the Otago Excursion Train along the Otago Central Railway, was closed by the New Zealand Railways Corporation in 1989 due to the lack of freight traveling through the area. A year after, the City Council of Dunedin bought the abandoned tracks and reestablished passenger service along the route. The route's thirty-seven miles make it New Zealand's longest tourist train, crossing through spectacular scenery.

The Taieri Gorge Limited travels along the old Otago Central Railway, which was completed in 1921. Providing passenger and freight service between Wingatui and Cromwell, the 146-mile Otago Central Railway was closed after the nationalization of New Zealand's railway system in 1990. The tracks between Clyde and Middlebranch were demolished that year to be converted into the Otago Central Rail Trail, a ninety-mile bike trail. Because of the action taken by the City Council of Dunedin, the tracks that remained between Middlebranch and Dunedin were preserved and now provide passenger service along the route.

Leaving Dunedin and passing the historic island-platform station of Wingatui, the train passes through the Salisbury Tunnel, the longest on the line, and then crosses over the Mullocky Gully along the Wingatui Viaduct. The viaduct is a historic wrought-iron structure built at the same time as this branch's construction in 1887. After the Wingatui Viaduct, the train then follows the Taieri River passing numerous abandoned stations, two curved viaducts, and a road-rail bridge at Hindon Road. The train then climbs out of the gorge toward the terminus of Pukerangi, passing rocky outcrops and then crossing over another road-rail bridge at Sutton Creek.

The Taieri Gorge forms between the high plateau of Maniototo and the Taieri Plains. It is twenty-five miles long, sixteen of which provide a route for the Taieri Gorge Limited.

The Taieri River is the fourth longest river in New Zealand.

THE MAEKLONG RAILWAY

This forty-two mile long railway through central Thailand operates on two different sections, the Mahachai Line and the Ban Laem Line, between the cities of Bangkok and Samut Songkhram. The Ban Laem Line runs between the two port cities of Samut Songkhram and Samut Sakhon, while the Mahachai Line connects Samut Sakhon and Bangkok. The transfer necessary between the two lines isn't as easy as switching platforms, but requires a ferry ride across the Tha Chin River to make a transfer at the Samut Sakhon Junction. The railway is famous for the Maeklong Railway Market, which is an open-air market that surrounds the Maeklong Railway Station in Samut Songkhram. The market is known as *Talat Rom Hup*, which translates to "umbrella pulldown market," and is notorious for the close quarters the market shares with the railway. Blaring its horn at deafening decibels, the train slowly works its way out of the Maeklong Station as vendors close their shop entryways as the train passes.

Built in two stages, the Mahachai Line was constructed by Tachin Railway Ltd. in 1904, while the Ban Laem Line was constructed by the Maeklong Railway Company in 1905. Both railways merged into the Maeklong Railway in 1907, which operated as a freight service that transported produce and fish from the Gulf of Thailand to Bangkok until 1942. During World War II, the railway was controlled by the military and was officially nationalized in 1946. Since then, it has been under the operation of the State Railway of Thailand, the state-owned railway system that transports nearly 35 million passengers a year.

Experienced vendors are able to place their product right next to the track without it being crushed as the train passes.

When the train isn't passing, the tracks second as a pedestrian path through the market.

The Maeklong Station has become a very popular tourist attraction.

THE BURMA RAILWAY

Hundreds of thousands of people were forced into labor to build the Burma Railway in just over a year.

This Thai railway that once connected the region of Ban Pong, a suburb of Bangkok, to the city of Thanbyuzayat, Burma, has a dark history of brutal imperialism. Nicknamed the "Death Railway," the Burma Railway was built by the Empire of Japan in 1943 to support its troops in World War II during the Burma Campaign. With the use of Thai and Southeast Asian populations and prisoners of war as forced laborers, the Empire of Japan subjected its workers to starvation, sickness, and harsh treatment as they constructed the 258-mile route. Estimates by the Australian government state that nearly 90,000 of the 330,000 laborers died while working on the route, with 16,000 of those who died being Allied soldiers. Although most of the track was closed just years after the end of World War II, a small section of the original track is still open today between Nong Pla Duk and Nam Tok, Thailand.

As Japanese forces invaded Burma in 1942, the Empire of Japan needed a good route to transport troops and supplies to the front without having to go around the long Malay Peninsula through the Strait of Malacca. Plans for the railway began soon after they entered the region, and construction began in September of 1942 with a projected end date of December 1943. Construction ended ahead of schedule in October of 1943, which surely contributed to the tens of thousands of lives that were lost. The line transported nearly 500,000 tons of supplies for the Japanese forces. By 1946, British troops ordered Japanese POWs to tear up the tracks between Nikki and Sonk rai, Burma, with the rest of the Burma track being deconstructed in phases.

The line officially closed in 1947, but a small seventy-three mile portion of track between Nam Tok and Nong Pla Duk, Thailand, was opened to passenger and freight service in 1957. The history of the railway has not been forgotten. Along various portions of the original track's route, there are memorials to remember the lives that were lost in the building of this short-lived railway. After the war, 111 Japanese military leaders were convicted of committing war crimes toward the POWs constructing the railway. Also, a particularly difficult section of the route, Hellfire Pass, has been turned into a museum and historic site to commemorate the lives lost at the site due to beatings, cholera, starvation, and exhaustion.

Nearly sixty-nine miles of track were originally laid in Burma, but all of the track that once ran through the country has since been deconstructed.

The sheer numbers of the Burma Railway are astounding. The 258 miles completed in such a short time, the 600 bridges built along the route, and the more than a quarter-million people who constructed the railway all make this railway an infrastructure project that will not be easily forgotten.

The North-South Train approaching the Hanoi Station along the famous Train Street.

THE NORTH-SOUTH RAILWAY

The North-South Railway, or the Reunification Railway, runs between the two former capital cities of North and South Vietnam, Hanoi in the north and Ho Chi Minh City in the south. Traveling 1,072 miles between the cities, it serves 278 stations, 191 of which border the North-South line that used to separate the two countries. Due to the country's history of colonialism in the nineteenth century and war in the twentieth century, the railway's infrastructure is in poor condition from bombings, sabotage, flooding, and deterioration. This poor condition has left the railway prone to a high rate of accidents, although the safety of the line has increased with recent investments into the infrastructure by development assistance. Nearly eighty-five percent of the Vietnamese Railway Network's passengers are served by the line, as well as sixty-percent of its cargo.

The North-South Railway took nearly thirty years to be completed. Built by the colonial government of France, construction of the railway began in 1899, working its way south from Hanoi section by section between stations, until it finally reached Ho Chi Minh City (then called Saigon) in 1936. After its completion, it took nearly sixty hours to travel the route from end to end, and that number was reduced to forty hours a few years later. The Japanese took control of the railway during World War II, which led to its extensive bombing by Allied Forces.

The Vietnam War led to even more bombing of the northern section of the railway by the Americans. After the Fall of Saigon and evacuation of American forces from South Vietnam, the new united Vietnamese state began to restore the railway, and the railway's 1,072-miles of track were back in operation by December 31, 1976.

The extensive railway consists of nearly 1,300 bridges—about 92,000 feet in total length—and twenty-seven tunnels that connect this once divided nation. Since its restoration in 1975, the tracks have begun to deteriorate quickly. Reduced speeds are required in tunnels due to weak infrastructure, while violent flooding tends to wreak havoc on the tracks passing through the center of the country. But despite the railway's tumultuous history, it provides passengers with hidden views of Vietnam's beautiful and lush environment. The train travels along the coastline of the country through numerous provinces. It also passes the scenic hotspots of the mountainous Hai Van Pass, the Lang Co Peninsula, the Van Phong Bay, and Hanoi's Train Street.

Top right: *The North-South Railway passing along an inlet of the East Vietnam Sea.*

Bottom left: *The train passing near the Hai Van Pass in the Annamite Mountain range.*

THE OIGAWA MAIN LINE

Japanese cherry blossoms are always a welcome sight along any train route. Operating here is one of the railway's electric multiple unit locomotives.

Connecting the Kanaya Station of Shimada to the Senzu Station of Kawanehon in Shizouka Prefecture on Japan's main island of Honshu, the Oigawa Main Line is operated by the Oigawa Railway company as a sightseeing tourist train. The Oigawa Main Line travels from the Shida Plains where Senzu is located up into the isolated regions of the Japanese Alps. The route totals twenty-nine miles in length along nineteen stations. Passangers frequently use the railway, which weaves around the Oi River, to reach hiking trails in the mountains or to access many of the natural hot springs in the area.

The Oigawa Electric Company built the railway in order to transport equipment and men up the Oi River to the construction site of the Nagashima Dam. Track construction began in 1927 from the Kenaya Station and reached the Senzu Station in 1931. Freight service was the exclusive cargo of the train until 1949, when passenger service began along with the electrification of the line. Freight service ended in 1983, but the railway has remained popular with tourists since then.

The region the Oigawa Railway passes through has a very low population density with no towns or cities along the route—many of the stations that the railway serves are unmanned island platforms. The railway uses both electric and heritage steam locomotives for its service. In 2016, the railway bought four new coaches in order to better preserve the historic coaches being used on the track.

Above: *The Oigawa Main Line operating one of their steam locomotives.*

Below: *The railway was built to ship materials to the construction site of this dam, the Nagashima Dam.*

A diesel locomotive powering over a large viaduct.

THE MAIN LINE

The Main Line in Sri Lanka connects the cities of Colombo on the Lacadive Sea of Sri Lanka's west coast, through the hill country of inner Sri Lanka, to the city of Badulla on 143 miles of track. Although there are plans to electrify sections of the track located within Colombo, the railway operates with diesel locomotives and dated lock-and-block signaling systems. Travelling east out of Colombo and its bustling suburbs of Ragama and Gampaha, the train begins an ascent near Rambukkana through the hill country's luscious orchards of tea gardens and local flora. Sheer cliffs line the tracks as the train continues to ascend between the stations at Balana and Kadugannawa. Eventually the climb gives way to views of forest and agrarian hillsides before the descent toward the terminus of Badulla.

The Main Line was built in sections starting from Colombo in 1864, following a nineteenth-century British route through the mountains. Working east of Colombo to Ambepussa, this section of track was the first railway to be built in the country, and service on the branch began in 1865. Throughout the following decades, the Main Line was extended further and further east in to the hills until it reached Badulla in 1924. Steam locomotives chugged in and out of the forest to ship tea and coffee from the hills to the port city of Colombo. Steam locomotives were replaced by diesel locomotives in 1953, and updates to the rolling stock came in 2011 with diesel-multiple-unit locomotives especially built to deal with the tracks steep gradients.

The Main Line is a broad-gauge track with a width of five and a half feet—an unusual size for a railway. It also gains quite a bit of elevation for a standard adhesion railway, climbing from near sea level to almost 6,300 feet above sea level at its highest point near the summit of Pattipola. The route is operated by Sri Lanka Railways with a variety of services provided to passengers. There are premium packages through ExpoRail, Rajadhani Express, or several other providers that will take you past the island's British-style tea gardens, mountains, valleys, and grand waterfalls.

Top right: *A Class S12 locomotive travelling the Main Line. The S12s were built for Sri Lanka Railways in 2011 by China's CSR Corporation.*

Bottom left: *Hiking along the Main Line has become popular, but it is not recommended.*

THE PEAK TRAM

The Peak Tram rising above Hong Kong's massive skyline.

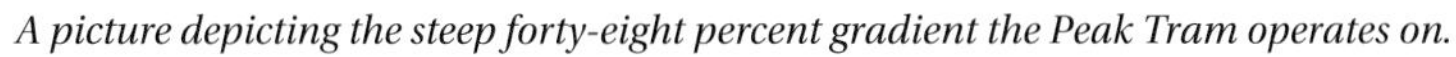

A picture depicting the steep forty-eight percent gradient the Peak Tram operates on.

The tram may look like it travels through isolated areas, but Hong Kong is the world's fourth most densely populated city.

Owned and operated by Hongkong and Shanghai Hotels, the Peak Tram is a residential and tourist funicular railway on Hong Kong Island. It travels no more than a mile but ascends nearly 1,500 feet from the Admiralty neighborhood of Hong Kong to Victoria Peak, Hong Kong's highest peak. Through the Mid-Levels of Hong Kong, an affluent neighborhood between Victoria Peak and the central business district of the city, the train is an exclusive opportunity to the see the hilly and metropolitan expanses of Hong Kong. The lower terminus of Garden Road Station is found near one of the oldest Anglican churches in China, St. John's Cathedral, in the twenty-two story St. John's Building. The upper terminus at Victoria Gap, hundreds of feet from the summit of Victoria Peak, is located within the luxurious shopping and entertainment complex called The Peak. There are four stations served between the two terminal stations.

Alexander Findlay Smith conceived of the Peak Tram in 1881. He received a concession for the railway just a few years later. Construction began in 1885 with the workers having to haul the equipment and material up themselves because of the area's inaccessibly steep gradients. The line was opened in 1888 and was a major factor in the development of Victoria Peak and the Mid-Levels. The service was originally built exclusively to be used by the residents of Victoria Peak, and between 1888 and 1926 service was segregated by class: first class serving British colonial officials and residents, second class for British military personnel and Hong Kong police, and third class for other people and animals. It was originally viewed as an engineering marvel at the time it was built. It operated with a steam engine until an electric motor was installed in 1926.

The line was completely renovated by a Swiss company in 1989, replacing the old track and adding a computerized control system. In recent years, the railway provides service to thousands of people a day, and in 2007, it was reported that the annual ridership was near 4 million people. Each train has a capacity of 120 passengers and can travel from end to end in under five minutes.

THE DAIKOFTO-KALAVRYTA RAILWAY

According to legend, the Vouraikos Gorge was home to a cave that was dedicated to Hercules. Pilgrims used to travel to the cave in order to have their fates revealed on the Tables of Knowledge.

This thirteen-mile journey through Greece's Vouraikos Gorge in the northern part of the Peloponnese Peninsula is a spectacular way to see the county's beautiful mountains and countryside. The Daikofto-Kalavryta Railway travels south from the city of Daikofto, on the coast of the Gulf of Corinth, to the mountain town of Kalavryta, which is situated among the Erymanthos and Aroania Mountains. Along the route, the train passes through the Vouraikos Gorge, named after Boura, the mythic daughter of Ion and Helice who was loved by Hercules, who supposedly opened the gorge to be close to her. Travelling out of the gorge, the train then goes through the town of Zachloro, which has a population of thrity-eight people and used to be the site of the Mega Spilaio Monastery.

The railway was built by Piraeus, Athens & Peloponesse Railways, which was founded in 1882 and nationalised into the Hellenic State Railway system in 1962. The railway company operated a number of railway routes through the country, but the Daikofto-Kalavryte Railway was the only railway in its system that was built on a smaller gauge than the rest. New diesel locomotives serve the seven stations along the line, although the railway still owns six original steam locomotives that were built for the railway in 1891.

The railway is a single-track railway that climbs from sea-level to near 2,500 feet above sea level. With a maximum gradient of seventeen percent, the Daikofto-Kalavryte Railway uses an Abt rack system for three sections of the route. As it passes through the Vouraikos Gorge, it crosses the Vouraikos River numerous times, travels through tunnels, and uses four passing loops. There are departures from each end several times a day, with each direction, taking about an hour from end to end.

Above: *Numerous built-out platforms, bridges, and tunnels were needed to complete the railway through the gorge's tenuous terrain.*

Left: *Although snow is not common in the Achaea region of Greece, it is very common in the Erymanthos Mountains, which have a higher altitude than other parts of the region.*

The three motor coaches that operate on the line are named Anne, Marie, and Jeanne, after the three daughters of the owner of the tramway when it was electrified.

THE MONT BLANC TRAMWAY

The Mont Blanc Tramway is a mountain railway that connects the small communal town of Saint-Gervais-les-Bains in the Haute-Savoie department of eastern France to the Mont Blanc massif. The railway operates on both standard adhesion and rack systems to traverse the near eight-mile ascent to the base of the Bionnassay Glacier. Nestled in the valleys and gorges of the Alps, Saint-Gervais-les-Bains occupies a wide range of altitudes from near 2,000 to 16,000 feet above sea level in its administrative area. The Mont Blanc Railway overcomes 5,879 feet of that difference between the Le Fayet Station at 1,903 feet to the Nid d'Aigle Station at 7,782 feet. The railway provides passengers with stunning views of Mont Blanc, the highest mountain of the Alps and all of the European Union at 15,777 feet, and the Mont Blanc massif, which consists of eleven distinct peaks all taller than 13,000 feet.

The original route of the Mont Blanc Tramway called for the railway to make it all the way to Aiguille de Gouter, a 12,674 foot peak in the Mont Blanc massif. The first section along the near five-mile route to the Col-de-Voza Station opened in 1907. The second section, which reached the railway's current terminus of Nid d'Aigle, opened in 1914. Unfortunately the original plans to extend the railway further into the massif had to be abandoned due to the outbreak of World War I. Construction never resumed, but the railway replaced its steam locomotives when it electrified the system in 1956.

The line uses its rack system for nearly eighty-five percent of the trip, relying on adhesion traction only at the foot of the line and intermediate stations. It has an average gradient of fifteen percent and reaches a maximum gradient of twenty-four percent. The Mont Blanc Tramway is very popular among hikers and mountaineers, who use the service for the access it provides to various trails in the mountain range. The terminal station of Nid d'Aigle provides access to the first stage of the most popular trail used to ascend the peak of Mont Blanc, the Gouter Route. In 2010, the track between the penultimate station of Mont-Lachat and Nid d'Aigle was closed for the rest of the season due to the risk of pooled glacial melt water flooding the area.

Passengers can transfer to the larger network of France's railway system at the junction at Le Fayet Station.

The Mont Blanc Tramway at the Col-de-Voza Station.

THE MONTENVERS RAILWAY

The Montenvers Railway approaching the Hotel de Montenvers.

Owned and operated by the same company that owns the Mont Blanc Tramway (see pages 140–141), the Montenvers Railway is a rack railway that connects France's rail network to the Hotel de Montenvers Station at the Mer de Glace, a valley glacier located in the Mont Blanc massif of the Haute-Savoie department of France. Departing from Chamonix, the site of the first Winter Olympics in 1924, the railway ascends up the sides of the Aiguilles de Chamonix up to an altitude of 6,276 feet, providing views of the massive Mer de Glace and the peaks of Les Drus (12,316 feet) and Les Grands Jorasses (13,806 feet).

The railway opened in 1908, operating on both rack and adhesion traction systems to climb the 2,858 foot difference between end to end. The average gradient is eleven percent and reaches a maximum gradient of twenty-two percent. The steam locomotives that operated on the track were replaced in the 1950s after the railway was electrified. Today, the railway has six electric railcars and three diesel locomotives. From the Hotel de Montenvers at the top of the line, passengers can enjoy a lovely meal at the panoramic restaurant, take a small cable car to a man-made ice grotto carved in the glacier, or explore the glacier museum in the hotel.

Above: *The Montenvers Railway ascending the rack railway toward the Mer de Glace.*

Left: *The Mont Blanc massif is covered by many glaciers, including the two longest glaciers in France, the Mer de Glace and the Miage Glacier.*

89
1221
52
6664 YDM4
嵯峨野
400
481